Hans-Florian Geerdes

UMTS Radio Network Planning:
Mastering Cell Coupling for Capacity Optimization

VIEWEG+TEUBNER RESEARCH

Advanced Studies Mobile Research Center Bremen

Das Mobile Research Center Bremen (MRC) erforscht, entwickelt und erprobt in enger Zusammenarbeit mit der Wirtschaft mobile Informatik-, Informations- und Kommunikationstechnologien. Als Forschungs- und Transferinstitut des Landes Bremen vernetzt und koordiniert das MRC hochschulübergreifend eine Vielzahl von Arbeitsgruppen, die sich mit der Entwicklung und Anwendung mobiler Lösungen beschäftigen. Die Reihe „Advanced Studies" präsentiert ausgewählte hervorragende Arbeitsergebnisse aus der Forschungstätigkeit der Mitglieder des MRC.

In close collaboration with the industry, the Mobile Research Center Bremen (MRC) investigates, develops and tests mobile computing, information and communication technologies. This research association from the state of Bremen links together and coordinates a multiplicity of research teams from different universities and institutions, which are concerned with the development and application of mobile solutions. The series "Advanced Studies" presents a selection of outstanding results of MRC's research projects.

Hans-Florian Geerdes

UMTS Radio Network Planning: Mastering Cell Coupling for Capacity Optimization

VIEWEG+TEUBNER RESEARCH

Bibliographic information published by the Deutsche Nationalbibliothek
The Deutsche Nationalbibliothek lists this publication in the Deutsche Nationalbibliografie; detailed bibliographic data are available in the Internet at http://dnb.d-nb.de.

Dissertation Technische Universität Berlin, 2008

D 83

Gedruckt mit freundlicher Unterstützung des
MRC Mobile Research Center der Universität Bremen

Printed with friendly support of
MRC Mobile Research Center, Universität Bremen

1st Edition 2008

Readers: Christel A. Roß

Vieweg+Teubner is part of the specialist publishing group Springer Science+Business Media.
www.viewegteubner.de

Cover design: KünkelLopka Medienentwicklung, Heidelberg
Printing company: STRAUSS GmbH, Mörlenbach
Printed on acid-free paper
Printed in Germany

ISBN 978-3-8348-0697-0

Abstract

This thesis deals with UMTS radio network planning, which aims at achieving maximum coverage and capacity in third-generation cellular systems at low cost. UMTS uses W-CDMA technology on the radio interface. More traffic can be served than in previous systems, but the precise capacity of a cell depends on the current user positions, traffic demands, and channel conditions. Furthermore, cells are coupled through interference and need to be considered jointly. Static system models include all these factors and support network planning decisions in practice. They contain thousands of users in typical city-wide scenarios and model each link explicitly. Until now, only time-consuming simulation methods are known for accurately evaluating expected network capacity under random variations of the input data. Known optimization schemes either rely on a simplified capacity model, or they use complex models, which are hard to analyze theoretically.

These problems are addressed with new models and methods for UMTS capacity evaluation and planning. Interference-coupling complementarity systems are first introduced as a concise static system model. They extend known models by including the new concept of perfect load control. This allows to treat individual users implicitly even if some traffic cannot be served, so the model dimension depends only on the number of cells. Subsequently, expected-coupling estimates provide a first-order approximation of expected cell load and user blocking. Their computation requires no simulation, but only a single evaluation of the complementarity system. Experiments on realistic data validate the new system model and confirm that the performance estimates are informative for planning.

The expected-coupling estimates are the basis of a new challenging optimization model that maximizes expected capacity via a deterministic objective. The model contains an accurate notion of cell coupling and is stated in closed form. It admits structural analysis, which leads to new mixed integer programming formulations, lower bounds, and heuristic solution algorithms. Four case studies on large realistic datasets demonstrate that the planning heuristics run efficiently and produce highly efficient configurations.

The results establish a top-level perspective on the relations between cells in UMTS radio networks. This paradigm allows new insights into capacity optimization and makes effective radio network planning with an accurate notion of capacity computationally feasible in practice.

Acknowledgments

My scientific development was only possible with the help and support of many people. I thank Andreas Eisenblätter for invaluable guidance, advice, and discussions. I thank my professor, Martin Grötschel, for letting me work in a challenging and stimulating environment, in which I never lacked for anything. My work has greatly profited from many discussions with colleagues in the graduate school MAGSI at TU Berlin, in the research center MATHEON in Berlin, in the MOMENTUM project consortium, in the COST 273 sub working group MORANS, and with my colleagues at the Optimization Department at ZIB. I thank the team at the ZIB library for always quickly providing me with any literature I needed. I thank my students Normen Rochau, Franziska Ryll, Christian Schäfer, and Jonas Schweiger for their help in experiments and for insightful discussions. I also thank Prof. Adam Wolisz for agreeing to referee this thesis.

This thesis could not have reached its final version without the help of a number of people. I thank Mathias Bohge, Andreas Eisenblätter, Marc Pfetsch, James Groß, Tobias Harks, Bertolt Meyer, Christian Raack, Jonas Schweiger, Hans Selge, and Ulrich Türke for careful and quick proof-reading and their constructive feedback. I am indebted to Mario Olszinski for his out-of-the-box thinking and ingenious tips in questions of design and layout. I thank Andreas Eisenblätter for ceding and adapting his implementations of crucial software components.

I thank my parents for supporting me throughout the years; without it, this all would have been impossible. I thank my twin sister for being there, always and before everybody else. I thank my grandmother for accommodating me in the comfort of her home for three weeks, during which I found the peace to develop many ideas that shaped this work. I thank my friends for distracting me. I thank Martin for his enduring support.

Acknowledgements

Contents

Introduction

The *universal mobile telecommunications system* (UMTS) is a technical standard for a third generation (3G) telecommunication system. UMTS provides data rates more than three times higher than its second generation (2G) predecessors.[1] The increased speed enables, for example, video calls, music downloads, or fast web surfing. The technology is already widely available. In 2007, a total of 166 commercial UMTS radio networks are operational in 66 countries covering all continents (UMTS *Forum*, 2007), and there are already more than 100 million UMTS subscribers (*3GToday.com*, 2007; UMTS *Forum*, 2006). The market for mobile telecommunication is, however, increasingly competitive, therefore operators need to invest effectively. In Western Europe, mobile phone penetration has reached 100 % in 2006 and the average revenue per customer is declining (*3G.co.uk*, 2007). Besides spectrum license fees, the main cost driver is network infrastructure (*Ellinger et al.*, 2002). Radio network planning can cut operational and capital expenditure by up to 30 % (*Dehghan*, 2005).

Good radio network planning is difficult for UMTS, because its radio interface is more complex than anything used at mass-market level before (*Dehghan*, 2005). Connections are separated via codes; they share the same frequency band and are thus subject to *interference*. On each link, the system constantly regulates the amount of generated interference to a minimum via a power-control feedback loop. Reducing interference via power control increases capacity, but the resource utilization changes dynamically and depends on the position of the users, their resource demand, their transmit activity, etc. Any capacity estimate that is sufficiently accurate for long-term radio network planning needs to consider the impact of all these factors on the power-control mechanism, which operates on a timescale of a few milliseconds.

In consequence, operators face new challenges in practical radio network plan-

[1]Realistic data rates of 144 kbps for UMTS vs. 40 kbps for GPRS (*Ellinger et al.*, 2002); with the HSPA extensions, UMTS is even faster.

Figure 1.1: Network planning in the life cycle of a radio network; the focus of this thesis is highlighted in boldface.

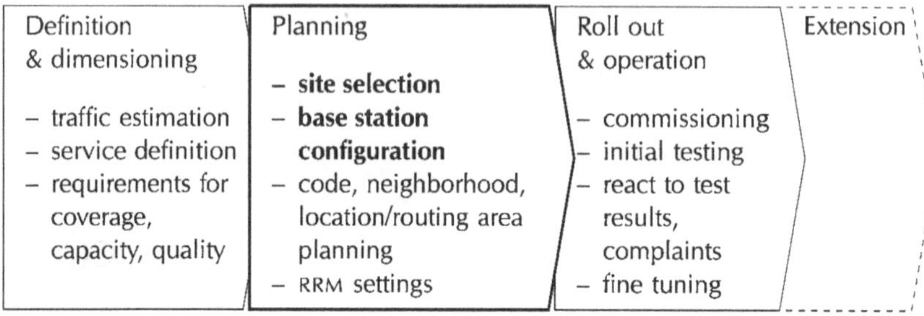

Definition & dimensioning	Planning	Roll out & operation	Extension
– traffic estimation – service definition – requirements for coverage, capacity, quality	**– site selection** **– base station configuration** – code, neighborhood, location/routing area planning – RRM settings	– commissioning – initial testing – react to test results, complaints – fine tuning	

ning. Before, the capacity of a radio cell could be determined independently from its neighbor cells; coverage and capacity planning were separated and treated sequentially; users accessed a standard voice service with a fixed resource requirement; and simple closed-form solutions to stochastic phenomena were at hand. Now, cells are coupled through interference, which necessitates a joint view on the interrelation of hundreds of cells; coverage is limited through interference, so coverage and capacity must be considered simultaneously; the impact of a user depends on the accessed service; and simulating the interplay of random user demand and network mechanisms is the only known method allowing an informative view on performance.

This thesis addresses the new challenges posed by UMTS technology. It contributes to a better understanding of third generation radio networks and develops new models and methods for planning them.

Scope. Our focus is the planning conducted prior to network roll out and operation. The main deployment phases of a radio network are sketched in Fig. 1.1 (simplified from *Nawrocki et al.*, 2006, Ch. 10). A preceding initial definition and dimensioning stage produces a business case; planning aims at implementing it in a concrete setting specified by detailed data on user demand, signal propagation, and equipment characteristics. Later adjustments are expensive, so a smooth operation has to be ensured during planning.

Network planners choose base station sites and configure antenna and cell parameters such that maximum coverage and capacity are attained with low cost. The decisions in large-scale network planning commonly rely on static simulation techniques; these allow sufficiently accurate performance analysis with reasonable computational effort. Static models describe network performance for a given, fixed configuration of served users called a *snapshot*. The dynamics of power control are handled by assuming *perfect power control* and averaging over a short time. Expected coverage and capacity is determined over a distribution on snapshots. This thesis aims at optimizing network performance as determined by static performance analysis.

Contributions beyond the state of the art. The relevant previous scientific work divides into three categories: theoretical analysis of interference-limited systems, practice-oriented system modeling and performance evaluation, and automatic performance optimization. In theoretical analysis, the power-assignment problem has been treated algorithmically and the capacity regions of single cells have been described. System modeling has refined and calibrated static models to faithfully represent a UMTS system; accurate performance evaluation uses static models in Monte Carlo simulations. In addition, analytical approximations for network performance have been developed. In automatic performance optimization, the trade-off between cost and coverage has been investigated extensively; capacity optimization until now mainly embeds static evaluation into simple heuristics, but eludes mathematical optimization.

We first contribute the new notion of *perfect load control.* This extends the classical static model to cases, in which a cell cannot serve all users in a snapshot. By combining perfect load control with perfect power control, we generalize the known interference-coupling equation system to *complementarity systems.* These describe user blocking and how power control in neighboring cells reacts to it. Individual users are only treated implicitly, so the complementarity systems allow a comprehensive definition of network capacity at cell level, *i. e.*, without considering individual links. The model reduces simulation effort for individual snapshots. Furthermore, the complementarity systems enable *expected interference coupling* estimates, which approximate simulation results with deterministic methods. We thus provide a means for quickly assessing cell capacity with the accuracy that network planning requires. Related developments have been made in parallel by other researchers; we extend them by estimates for user blocking, we add new analytical tools for assessing the method's accuracy, and extend it in critical parts. All contributions to modeling and performance evaluation are validated on realistic planning data through comparison with detailed simulation.

The advances in system modeling enable a novel closed-form *capacity optimization model* based on the expected interference-coupling matrix. The objective function is nonconvex; we optimize it with search methods and, alternatively, approximate it by a mixed integer linear program. Computational case studies on realistic datasets demonstrate that our methods have significant impact on network capacity as computed by detailed simulation. Analyses of the case-study results show that the configurations are efficient and that capacity optimization is satisfyingly addressed by our optimization methods.

Thesis structure. The three subsequent chapters treat performance modeling and evaluation. Ch. 2 reviews the basics of UMTS technology, introduces the classical static model, and discusses its use in performance analysis through simulation. In Ch. 3, perfect load-control is derived, which leads to the deduction of the interference-coupling complementarity systems as a model of performance on an individual snapshot. Ch. 4 develops deterministic methods, notably expected-coupling approximation, to obtain estimates for the expected performance subject to the snapshot distribution; this is followed by computational experiments validating our performance evaluation

Figure 1.2: Software infrastructure and data flow

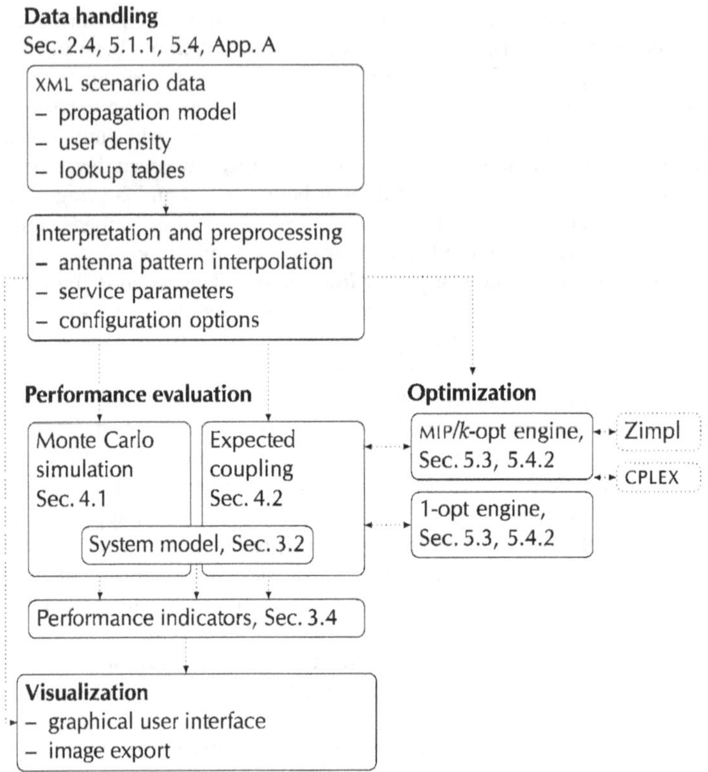

approach. Ch. 5 then treats the optimization of expected network performance. We survey the literature and describe our new models and approaches. They are tested in four case studies, whose results are further analyzed and discussed. We conclude and indicate perspectives beyond the present work in Ch. 6.

Implemented software. A set of software components has been implemented in JAVA™ for handling the planning data, performing computations, and generating all images of radio networks and their performance. Fig. 1.2 illustrates the different modules. The data used for all computations and images are four planning scenarios represented in an XML format[2] developed by the MOMENTUM project team and refined by the MORANS initiative with the author's active contribution. The raw data is accessed using the XML data binding framework Castor[3]. The data is preprocessed to a

[2]The format is documented at http://momentum.zib.de/schemata/docu/.

[3]The Castor Project, http://www.castor.org

representation suitable for the performance evaluation and optimization modules described in the following chapters. The Apache Commons Math library (*O'Brien*, 2004) is used for numerical integration and for generating random numbers. Mixed integer programming formulations are implemented with Zimpl (*Koch*, 2004) and solved with the CPLEX callable library (ILOG *S. A.*, 2006); the code for antenna pattern interpolation and the 1-opt engine are native C implementations kindly provided by Atesio GmbH. Input data and analysis results are visualized in a custom GUI and exported to vector graphics using the FreeHEP library (*Johnson & Youssef*, 2000).

Bibliographical note. Parts of this work have been recently published in collaboration with other researchers. The relevant previous publications are listed here; all third-party material (not co-authored by the author) is attributed where it is introduced. The idea of load scaling (Sec. 3.2.1) first appears in: Eisenblätter A, Geerdes HF, Koch T, Martin A, Wessäly R. UMTS *radio network evaluation and optimization beyond snapshots.* Math Methods Oper Res, 63, pp 1–29, 2006. The paper also contains an expected-coupling scheme and an early version of the mixed integer programming model for capacity optimization. It is refined in: Eisenblätter A, Geerdes HF. *A novel view on cell coverage and coupling for* UMTS *radio network evaluation and design.* In *Proc INOC'05*. ENOG, Lisbon, Portugal, 2005. The optimization framework and, in particular, the objective function, are new contributions. The contents of Sec. 3.2.3 are partly covered in: Eisenblätter A, Geerdes HF, Rochau N. *Analytical approximate load control in* W-CDMA *radio networks.* In *Proc IEEE VTC 2005 Fall*. IEEE, Dallas, TX, 2005*b*. Computational results along the lines of Sec. 4.4.1 also appear in the latter paper; they are elaborated here. The basic idea for the refined estimates of the grade of service as presented in Sec. 4.3 appears in: Geerdes HF, Ryll F. *Efficient approximation of blocking rates in* UMTS *radio networks.* In *Proc PGTS'06*. Wrocław, 2006. The derivation and motivation for the scheme is original to this work. The downlink transmit power part (3.40a) of the generalized pole equations, as well as partial computational results from Sec. 4.4.2 are published in: Geerdes HF, Eisenblätter A. *Reconciling theory and practice: A revised pole equation for* W-CDMA *cell powers.* In *Proc MSWIM'07*. Chania, Greece, 2007.

Data sources for computations and images. The basis for all images displaying planning scenarios and for computational experiments are datasets that have been provided by the consortium of the IST project MOMENTUM[4] and by the members of the sub-working group MORANS of the COST 273 initiative. For details, see Sec. 5.4.1.

[4]IST-2000-28088 MOMENTUM, http://momentum.zib.de

Radio network modeling and performance evaluation for UMTS

We review the foundations of radio communication and introduce the classical static approach to performance evaluation for UMTS; this provides the starting point for the subsequent developments. We first outline the general structure of cellular wireless communication networks and the special nature of the radio channel in Sec. 2.1. In Sec. 2.2, we explain the principles upon which communication over the UMTS radio interface is built, as well as their implementation in the standard. The available methods for UMTS radio network performance evaluation are briefly discussed in Sec. 2.3. We introduce the classical static model of the UMTS radio interface in Sec. 2.4; it is used in the remainder of this thesis. Its role in the evaluation of a expected radio network performance by static simulation is detailed in Sec. 2.5, which throws into relief the complexity of classical performance evaluation for UMTS.

Readers familiar with the topic might skip the remainder of this chapter and only study the notation used for our system model as summarized on p. 33 and in Definitions 1–3. A comprehensive list of notation is provided on pp. 169ff.

General literature. *Walke* (2001) gives an overview of cellular wireless technologies; an authoritative source on CDMA and spread spectrum communication is *Viterbi* (1995). Seminal monographs on UMTS planning are the books by *Laiho et al.* (2002) and *Holma & Toskala* (2001). A newer book describing many aspects in detail is (*Nawrocki, Aghvami, & Dohler*, 2006). We reference specialized literature as the occasion arises.

2.1 Cellular wireless communication networks

In cellular wireless networks, a fixed infrastructure installed by the network operator enables user communication. The forefront of this infrastructure is the *radio network,* a set of base stations with radio antennas. Users only communicate over radio with the fixed base station antennas, not among themselves.

2.1.1 Network infrastructure

Further network elements besides the antennas are needed to forward radio communication to the core network. Fig. 2.1 gives a schematic overview of the UMTS radio access network according to *Walke* (2001). It consists of many *radio network subsystems,* which are connected to the core network. The core network forwards calls and connections between its subsystems and to other networks, like the fixed-line network or the Internet. Exactly one *radio network controller* (RNC) controls each radio network subsystem. The RNC is responsible for signal processing, encryption, handover control, mobility management, and communication with the core network. Data is exchanged with the core network via the *mobile switching center* (MSC) handling circuit-switched traffic such as voice connections, or via the *serving* GPRS *support node* (SGSN) handling packet-switched data traffic.

An RNC controls one or more *Node-Bs.* A Node-B consists of a cabinet with the actual radio frequency transmitters and receivers and signal processing hardware. We will focus on the connection between the user equipment and the Node-B. If data is transmitted from the Node-B to the user, the communication is said to take place in the *downlink* direction (DL); the reverse direction is the *uplink* (UL).

The Node-B is connected to several antennas, through which it controls one or more *cells.* A cell is a logical organizational unit; its precise definition depends on the technology at hand. In our context, it is safe to picture a cell as a single antenna.[1] The area in which a cell serves users is its *cell area.* The antennas of *macro cells* are installed on rooftops or special antenna masts and serve a large area. In addition, there may be smaller cells served by micro cells or indoor pico cells.

2.1.2 The radio channel

A signal transmitted over a communication channel has to be clearly detectable to be properly received and decoded. The *signal to interference ratio,* the ratio of the desired signal's strength over all interfering signals and noise, measures the quality of the received signal. The signal to interference ratio has to exceed a threshold:

$$\frac{\text{Received Signal}}{\text{Noise} + \text{Interference}} \geq \text{Threshold}\,. \tag{2.1}$$

The threshold depends on the channel, the supported data rate, the coding scheme, the receiver hardware, and the application requirements. Simultaneously satisfying

[1] In practice, a cell may be served by several antennas. For the model developed in this chapter and used throughout this thesis, a cell is characterized by a common pilot channel and a single budget in several resource dimensions, such as orthogonal spreading codes and transmit power.

Figure 2.1: Structure of the UMTS radio access network

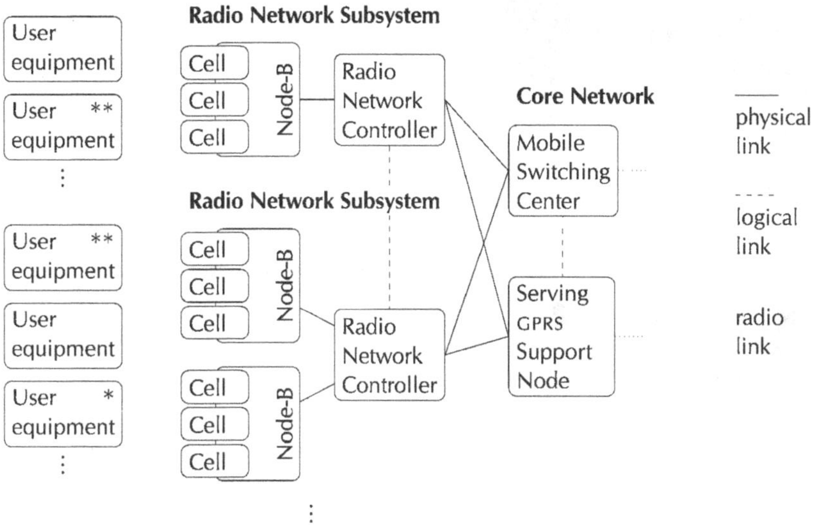

a constraint of type (2.1) for several users is one of the problems of communication theory and practice. A radio channel causes specific difficulties.

Interference. The first issue is related to the broadcast nature of the channel. In the case of cable connections, the parties that can take part in communication on the same channel are easily controllable by hardware. In the case of wireless communication, it cannot be avoided that a signal is propagated to other receivers than the intended one. There is a privacy-related side to this issue, which is addressed by encryption in modern digital communication systems. On the other hand, signals that arrive at a receiver for which they are not intended add to interference and therefore pose a problem for decoding the signal. We distinguish *co-channel* interference generated on the same frequency band and *adjacent channel* interference leaking in from neighboring bands.

Interference from third-party radio communication is largely ruled out in commercial mobile telecommunication because it takes place in licensed spectrum bands regulated by governmental bodies like the FCC in the US or the *Bundesnetzagentur* in Germany. The license entitles (and usually also obliges) the operator to exclusively use a certain part of the radio spectrum for wireless mobile communication. The spectrum allocated for UMTS comes in chunks of 5 MHz located mostly around 2 GHz.[2] The technology used on the radio interface needs to ensure that the expensive

[2]The main spectrum for W-CDMA FDD is allocated at 1920–1980 MHz (UL) and 2110–2170 MHz (DL) in Europe and most of Asia. In other parts of the world, the allocated frequencies vary; in the US,

Figure 2.2: Path loss function for an antenna in central Lisbon

exclusive frequency band is efficiently shared between users and that interference is avoided as far as necessary.

Attenuation. The second challenge for wireless communication is tied to its signal propagation properties. All communication channels attenuate a signal, but the attenuation per traversed distance is several orders of magnitudes higher for a radio signal than for cables, say. In vacuum, the strength of a radio wave decays proportionally to the squared distance between source and destination. On the earth's surface, the attenuation is higher. The inverse of attenuation is called the *channel gain*. In urban environments, several influences interact. The signal reaches the receiver on multiple distinct paths; on each path it is obstructed, reflected, diffracted, and refracted. Modern *rake receiver* technology (*Bottomley et al.*, 2006) partly remedies the multipath effects.

A precise account of all relevant influences is theoretically possible using Maxwell's equations, but this is practically infeasible because the necessary information cannot be obtained. Attenuation is therefore broken down into three components that are modeled separately (*Saunders*, 1999):

(a) *Path loss* denotes the median signal attenuation, which depends mainly on the distance. It is usually computed by a deterministic model that is adapted to frequency, density of surrounding building, etc. Popular models for urban areas are Okomura-Hata (for larger cells) and Walfish-Ikegami (for smaller cells); *Kürner* (1999) describes adaptations of these models to the frequencies used by 3G systems. If more specific data is available, ray-tracing and ray-launching methods produce best results (*Schmeink*, 2005, Pt. I). Fig. 2.2 illustrates the path loss calculated with an Okomura-Hata model; the antenna is indicated as a small arrow.[3]

notably, lower frequencies in the 824–960 MHz band occur (*Holma & Toskala*, 2001, Ch. 1).

[3]Configuration parameters (cf. Sec. 5.1.1): Kathrein K 742 265 antenna, azimuth 210°, antenna height

(b) *Shadow fading* describes the signal variation caused by obstacles on the transmission path. It is approximately stationary and often modeled as a random variable. For a moving user, the influence of obstacles varies on a time scale of several seconds; shadow fading is therefore also called *slow fading*. Models for shadow fading are discussed in Sec. 2.5.2.

(c) The short-term variations of the signal on a timescale of a few milliseconds are summarized under the term *fast fading*. They are mostly due to destructive and constructive interference between multiple paths of different length from sender to receiver. A typical fast fading pattern is shown in Fig. 2.5(a).

Positive quantities with a high dynamic range such as attenuation and power values are usually specified in *logarithmic scale*. The unitless ratio between two quantities $a > 0$ and $b > 0$ is stated in *decibel* (dB) as:

$$10 \cdot \log_{10}(a/b)\,\text{dB}\,.$$

Absolute power values are often specified in decibel over 1 mW (dBm); a power of a W corresponds to

$$(10 \cdot \log_{10} a + 30)\,\text{dBm}\,.$$

Formulas are consistently formulated in linear scale in this thesis.

2.2 The UMTS radio interface

The advantage of UMTS over 2G technologies stems largely from improvements on the radio interface; the core network has only been changed reluctantly, as it should remain compatible. On the radio interface, the standard has incorporated ideas and approaches to outperform second generation radio interfaces and address shortcomings of the previous systems. Besides plain speech telephony, for which 2G systems were originally conceived, 3G radio networks enable a variety of new services like video telephony, streaming, and packet-based data services like web browsing or e-mail. At the same time, radio spectrum is a scarce resource, and it has to be used sparingly and efficiently. The technology that has been chosen to address this demand is *wideband code division multiple access* (W-CDMA).

2.2.1 The principle of code division

Code division multiple access (CDMA) is one of several multi access schemes developed for sharing the capacity of a communication channel among several users. Other approaches to this problem are *time division multiple access* and *frequency division multiple access*, where time or the frequency band are subdivided into small units in which users can transmit exclusively. Both approaches are used, for example, in GSM. In CDMA, the dimension in which users are separated is code. The principle of code division multiplexing is illustrated in Fig. 2.4.

---

above ground 25 m, electrical tilt 6°, mechanical tilt 0°.

Figure 2.3: Code tree for constructing orthogonal codes of variable length.
The three marked codes can be used simultaneously.

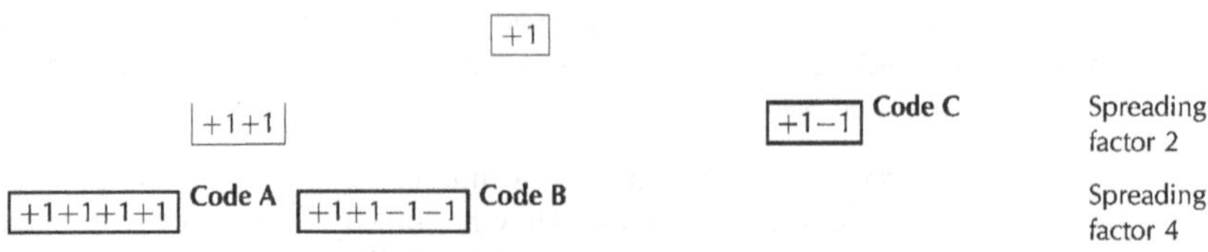

Coding. Each connection is assigned a (locally) unique code, *i. e.*, a sequence of symbols. A tree providing different codes is shown in Fig. 2.3. To encode the message, the transmitter multiplies each symbol to be transmitted with the code sequence. In the example we use binary modulation, so the transmitted sequence per bit is either the code sequence itself, if the user bit is $+1$, or its inverse, if the user bit is -1. The scheme is likewise used with more sophisticated modulations producing symbols from the complex (phase/amplitude) domain. The transmissions from all sending users superimpose on the channel, so all receivers in the system receive the sum of all encoded sequences.

The transmitted sequence has a higher rate than the original sequence. In W-CDMA, a piece of user information (before encoding) is called *bit* and a symbol from the encoded sequence is called *chip*. The codes can have different lengths, but the chip rate is constant for all users. In consequence, shorter codes enable a higher data-rate. Codes A and B in our example consist of four symbols, whereas code C has only two bits. User C can hence transmit twice as much data per time unit as users A and B.

The receiver uses the same code that encoded the information for decoding the received sequence of symbols and retrieves the original bit sequence. Each symbol is multiplied with the next element of the cyclically traversed code sequence. If all codes have been chosen appropriately (see below), the sum over the resulting sequence corresponds to the original user data and undesired signals are removed.

Spread spectrum and interference. W-CDMA is called a *spread spectrum* transmission scheme because the bandwidth of the transmitted signal is higher than that of the user data; this has advantages for suppressing interference from within the system and from third parties. The code's *spreading factor,* the number of symbols in the code, governs the degree of spreading. How exactly interference is reduced, depends on its characteristics and on the properties of the codes. In our simple example, the channel is ideal in that no noise occurs, and the codes are chosen such that interference from other users cancels out. In general, the reduction of unwanted signals is called the *processing gain;* it is equivalent to the spreading factor. With a higher spreading factor, a higher processing gain is achieved and interference is better suppressed, but the user bit rate is smaller. In our example, transmissions A and B have a processing gain of four, whereas in case C the gain is only two.

Figure 2.4: Example for code multiplexing with binary modulation using the codes from Fig. 2.3

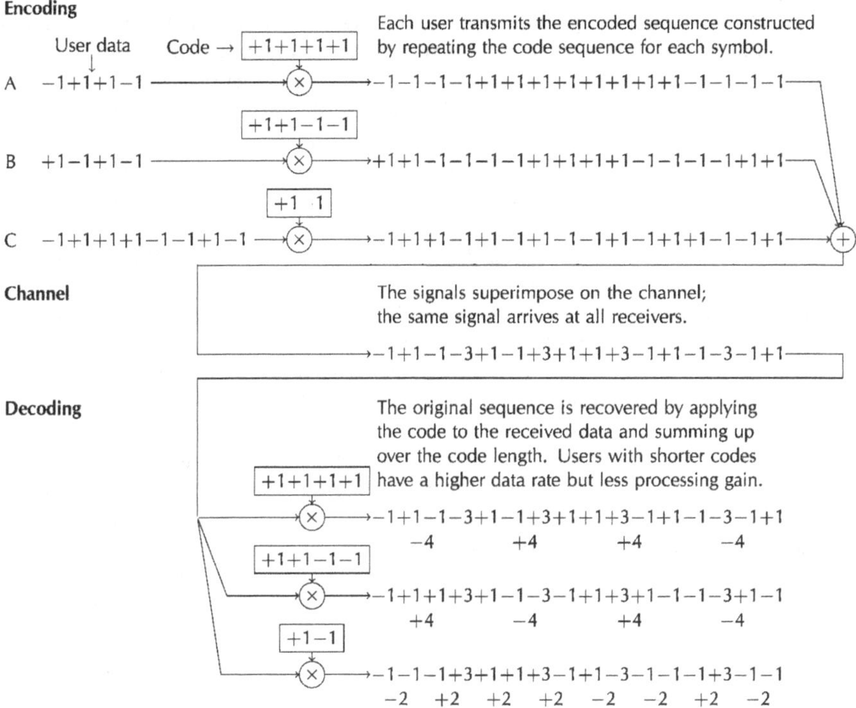

2.2.2 Wideband code division multiple access

The UMTS systems considered in this thesis employ *frequency division duplex* (FDD): separate radio bands are reserved for uplink and downlink. There might be several carriers, *i. e.*, associated pairs of uplink and downlink bands, but we will focus on single carrier systems in which all transmissions in one direction (uplink or downlink) are transmitted on the same frequency and create co-channel interference on each other. We neglect any mutual interference between uplink and downlink and other adjacent-carrier interference, so *interference* always means co-channel interference. In cellular systems, interference is composed of *intra-cell interference* stemming from transmitters in the same cell and *inter-cell interference* stemming from transmitters in other cells.

Orthogonal and pseudo-noise codes. Two kinds of codes are used in UMTS: orthogonal codes and pseudo-noise codes. Orthogonal codes cancel each other out if received in perfect synchronization and therefore do not create any mutual interference. Transmitters are, however, not synchronized in UMTS, so orthogonal codes can be used only to separate signals from a single source. Orthogonal codes of arbitrary length can be constructed from the code tree shown in Fig. 2.3; they are mutually compatible as long as no code in use is the ancestor (in the code tree) of any other. Different transmitters are distinguished via pseudo-noise sequences, which have statistical properties close to random sequences of bits. When decoding a message with a pseudo-noise sequence it was not encoded with, the result is virtually identical to decoding random noise. If the same code sequence is used and the receiver is synchronized, the message is recovered with the applicable processing gain.

Each transmitter hence consecutively applies two codes to the message: An orthogonal spreading code called the *channelization code* first increases the signal's rate to the chip rate. Subsequently, the pseudo-noise *scrambling code* is applied to enable separation of different transmitters. The scrambling code does not affect the rate of the signal any further.

This mechanism causes a fundamental difference between the uplink and downlink directions. All downlink transmissions within one cell originate at the same receiver, so they do not interfere as long as pairwise orthogonal codes are available. When a mobile decodes a signal, it is therefore only subject to interference from other cells, not from the same cell—at least in theory. In practice, the transmitted symbols partly overlap under multi-path propagation conditions and orthogonality is compromised. Besides, the code tree is occasionally exhausted and an additional secondary tree is used. We will take into account the loss of orthogonality, but not the capacity of the code tree. The problem of best satisfying incoming requests for new orthogonal codes in an on-line fashion is treated, for example, by *Chin et al.* (2007).

In the uplink direction, on the other hand the mobiles do not use codes that are mutually orthogonal, so the different signals are only distinguished by their different pseudo-noise sequences. In consequence, all arriving signals fully count as interference for a receiving Node-B, and there is no difference between interference from the own cell and interference from other cells. There is a large number of pseudo-noise sequences of reasonable length; it is therefore easy to ensure that the transmitters use distinct scrambling codes (*Laiho et al.*, 2002, Ch. 4).

Communication over the radio interface: pilot and dedicated channels. Based on physical radio transmission, the access network provides logical *transport channels* for exchanging user data with mobiles. For example, transport channels supporting data communication or voice traffic are available. Channels are further subdivided into *common channels* potentially accessed by all mobiles and *dedicated channels* providing point-to-point connections between a mobile and the Node-B. There are many different channel types, but classical static models for W-CDMA radio networks focus on one common signaling channel and one dedicated transport channel.

The *common pilot channel* (CPICH) is the control channel that governs the mobile's association to the network. To select a cell, the mobile scans all cells within reach, synchronizes to each, and decodes a predefined beacon signal emitted on the pilot channel. In this way, the mobile estimates the link quality to the Node-B; usually it attempts to connect to the cell providing the strongest pilot signal. Because it carries no user information, the signal *strength* of the pilot channel is measured as the received chip energy E_c. The signal *quality* is measured relative to the total noise and interference density I_0, it is therefore called E_c/I_0. For successful association to the network, both the strength and the quality of the pilot signal must be sufficient; if this is the case the mobile has *coverage*.

There is only one type of dedicated (transport) channel in uplink and downlink. For the time being, dedicated channels carry the vast majority of user data. This may change in the future, when high-speed packet access (HSPA) channels for transmitting packet data are more widely used. HSPA channels enhance the capabilites of the radio interface and the peak data rate by additional features and mechanisms (*Hedberg & Parkvall*, 2000; *Holma & Toskala*, 2006).

Dedicated channels use codes of various spreading factors and support different data-rates. For allowing a flexible definition of new services, logical *bearer* services are defined in UMTS. A bearer guarantees a certain quality of service between two points in the network in terms of data rate, error probability, delay, etc. For radio network planning, the radio bearer services between mobile terminal and RNC are relevant. Dedicated channels also provide a mechanism for varying the bitrate as necessary during a connection. In particular, the channel remains inactive when no data is to be transmitted. This is, for example, the case in voice communication when one person listens while the other one is speaking. Also in a web browsing session data is not transmitted continuously, but there are pauses, for example, between TCP packets.

The quality of a signal received on the dedicated channel is denoted by E_b/N_0. E_b stands for "bit energy", the level of the *decoded* signal, including the processing gain. N_0 denotes the "interference and noise power". Only in downlink, there is a difference between I_0 and N_0 as the latter takes into account the orthogonality of signals from the own cell. Closely related to the E_b/N_0 value is the *carrier to interference ratio* (CIR); they are equivalent up to the processing gain.

Soft Handover. As all transmissions take place in the same band and at the same time, a user may connect to more than one base station at a time; this is called *soft handover*. In uplink, different antennas receive the same signal, and either the RNC combines the decoded signals from different Node-Bs (soft handover) or one Node-B decodes the information using both undecoded sequences (softer handover). The latter leads to less decoding errors and is thus more advantageous. In Fig. 2.1, the mobiles marked with (∗) is in soft handover, while mobile (∗∗) is in softer handover. In downlink, the mobile receives signals from different antennas and decodes their combination.

Soft handover has several advantages: the handover process is more likely to be successful, as a new connection can be setup before breaking the old one. Resource requirements for users at cell edges are lowered, because the error probability is reduced due to macro-diversity. On the other hand, more than one cell has to spend resources on the user. Although analytic investigations calculate a small gain in the required E_b/N_0 (e. g., *Laiho et al.*, 2002, Ch. 4.6.1), the net effect measured in practice is reportedly a deterioration (*Nawrocki et al.*, 2006, Ch. 10.2.4.3), but coverage and stability are improved.

Figure 2.5: Fading and inner-loop power control

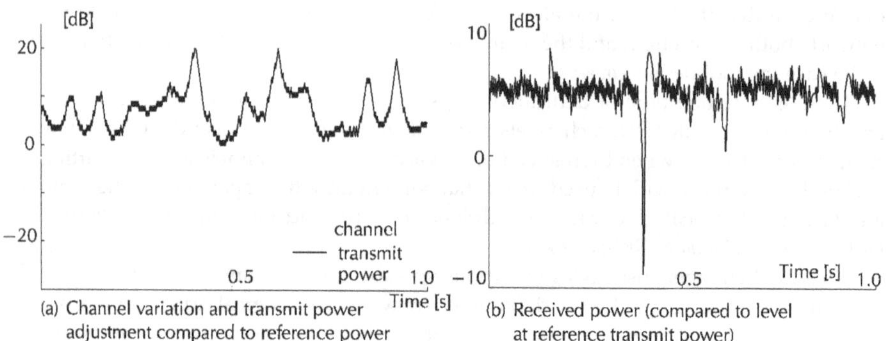

(a) Channel variation and transmit power adjustment compared to reference power

(b) Received power (compared to level at reference transmit power)

2.2.3 Power control

Besides coverage, interference is a critical limiting factor of W-CDMA networks. Several mechanisms for interference reduction have therefore been integrated into the standard, notably *closed loop power control* on dedicated channels. It adjusts the transmit power to the minimum level at which the receiver can properly decode the signal. Power control thus strikes the balance between providing sufficient signal quality for the receiver and avoiding interference for the system. The loop is considered closed, because the receiver gives immediate feedback to the sender.

Outer loop. The outer loop of power control adjusts the right-hand side threshold in (2.1) such that a bearer-specific target error rate is approximately met. Real-time applications like video-telephony require a low target error rate. Services that tolerate moderate delay, like web browsing, can handle errors with retransmits and are therefore less demanding. While data is being received on the link, the system compares the actually observed decoding error to the target value and uses the result to periodically adjust the threshold on the right-hand side of (2.1). If too many errors occur, the threshold is increased, and vice versa; updates occur 10–100 times per second.

Inner loop. The inner loop of power control adjusts the numerator in (2.1) such that the E_b/N_0 ratio meets the current threshold. With a frequency of 1 500 Hz, the receiver specifies a power-control bit on a signaling channel in the reverse direction. The transmitter accordingly increases or decreases the power by a fixed amount, usually 1 dB. (The power level cannot remain unchanged.) The update cycle is fast enough to timely react to fast fading in most cases.

Fig. 2.5 illustrates the functioning of the inner-loop power control mechanism over the course of one second.[4] The light line in Fig. 2.5(a) indicates the variation of the

[4]Channel: power envelope of a channel subject to Rayleigh fading, center frequency 2.4 GHz, user speed 1 m/s, uncorrelated scatterers. Power control: own simulation at 1 500 Hz, delay of two timeslots, error probability 5 %.

radio channel for a given connection over the course of one second in dB. The dark line indicates the value by which the transmit power is adjusted by power control; the plot implicitly assumes a constant level of interference at the receiver, so the transmit power basically mirrors the channel state. The resulting power level at the receiver is depicted in Fig. 2.5(b).

2.2.4 Congestion control

If too many users are present, the proper functioning of the radio network is put to risk and the system counteracts. Following the presentation of *Laiho et al.* (2002, Ch. 4.4), we group the functions of the network that are responsible for controlling the load on the radio interface under the term *congestion control*.[5] Congestion control is not specified in the UMTS standard, neither is terminology unified. The precise implementation is hence vendor-specific; it may even vary drastically with the firmware versions at the Node-B and at the RNC (*Mäder et al.*, 2006). Three mechanisms play a role for keeping the load in check:

(a) *Packet scheduling* handles packet data traffic. It decides, when non-real time traffic is delivered to the receivers, and with which data rate. If the load approaches the limit, ongoing transmissions may not increase their bitrate any further. If the system is in overload, bit rates may be decreased, connections on dedicated channels may be shifted to a common channel (and served on a best-effort basis only), and connections may eventually be dropped.

(b) *Admission control* decides if and how requests for radio access bearers are granted. To prevent an impending overload situation, admission control only admits a new user if the system is not pushed to overload. Users arriving in a handover process have priority over those trying to setup a new connection, as it is considered more annoying for a customer if a call is disrupted than if it fails to be set up. For packet connections, admission control regulates the supported connection speed. If the system is in overload, no new bearers are assigned.

(c) *Load control* actively reduces system load by sacrificing ongoing connections in a controlled fashion. This should only occur in exceptional cases, as under normal conditions packet scheduling and admission control avoid overload situations. The Node-B may influence power control (refraining from sending or acting upon "power up" commands in the inner loop; lowering the CIR target in the outer loop). The radio network controller may lower the bitrate of circuit switched services where possible, attempt to initiate the handover of users to a different carrier or system (e. g., GSM), and eventually drop single calls.

[5]The term "congestion control" is used differently in different contexts in telecommunications theory. We only use it to refer to the functionality of the radio interface as explained below. Our use is not to be confused with other uses in related scopes such as TCP's end-to-end congestion control mechanism.

2.3 Methodology of performance evaluation

Third generation radio networks are spectral-efficient because they adapt swiftly to changing conditions, but this poses difficulties for performance evaluation. Signal variations are equalized by the power control mechanism; interference is reduced to the minimum level; radio resources are only consumed by a user when data has to be sent; even the data rate and coding scheme is adapted on HSPA channels. Because of all these effects, UMTS radio networks have inherent dynamics. Network performance and capacity utilization are hence intricate to define; simulations can only determine them approximately. One distinguishes *dynamic* and *static* simulation methods.

Dynamic simulation provides the most accurate picture of W-CDMA radio networks by emulating all dynamic effects and the network's reaction in detail. The considered effects include fast fading of the radio channel, user movement over time, and the activity of data transmissions. Crucial network mechanisms are power control, handover, and congestion control. The simulation is performed event-driven or time-driven. A time-driven dynamic simulator for UMTS is, for example, described by *Fleuren et al.* (2003). To assess the stochastics of key performance indicators, their variations are observed over time (possibly after awaiting an initial transient phase). This has to be done over a simulated period of at least several minutes, so the process is computationally very expensive, and the software is difficult to implement.

Static simulation has been developed as an alternative that is capable of analyzing large radio networks with manageable effort and reasonable accuracy (*Wacker et al.*, 1999). The network's behavior is evaluated repeatedly on static representations of user configurations called *snapshots*. The concept is to average the effects varying on a short time span of roughly a few seconds. This includes the vacillation of transmit powers due to power control, fast fading on the radio channel, and short-term on-off activity patterns on a radio link. Quantities that change over times significantly larger than the observation time are taken into account by defining a random distribution over the snapshots. This includes shadow fading on the radio channel, user movement, initiation and termination of connections, and changes in connection speed. Snapshots are usually drawn independently, but some investigations require correlated sequences (*Du et al.*, 2004; *Lamers et al.*, 2003).

The main tool used in static simulation is a collection of equations and inequalities, which we refer to as a *static model* (*Hoshyar et al.*, 2003); different static models are evaluated many times during simulations following the Monte Carlo principle. A static model contains information on the network configuration, propagation conditions, and on the position and service demand of users in a snapshot. The model is used to determine whether the set of users can be simultaneously served by the network, and, if so, which average transmit powers are needed. The expected performance of a network configuration is determined by repeatedly evaluating the performance for snapshots drawn from the random distribution.

Static modeling and simulation is universally accepted and used in network planning theory and practice whenever more refined methods are too time-consuming. Examples for commercial network planning software employing static methods are *Forsk SA* (2007); *Lustig et al.* (2004); *Optimi, Inc* (2006*b*); besides, most scientific pa-

Figure 2.6: Users, antennas, links, and interference

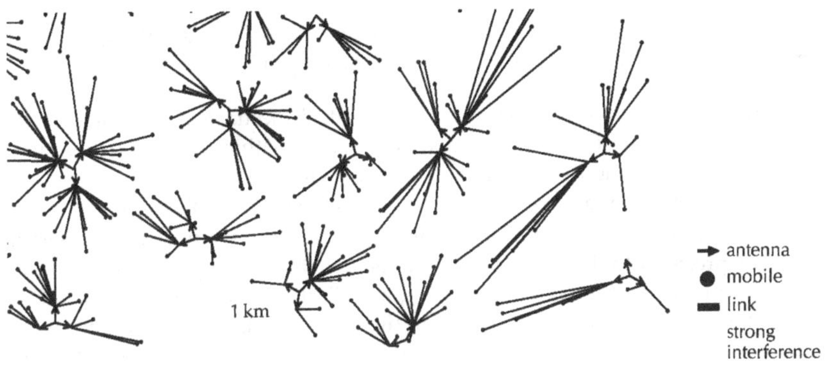

pers on UMTS cited below use static models. Dynamic simulators are chiefly used for investigating special aspects or for calibrating static models in small test scenarios (*Buddendick*, 2005); they are included in some commercial products (*Actix Inc*, 2006; *AWE Communications GmbH*, 2005*a*; *Optimi, Inc*, 2006*b*).

There are few publications on the general validity of static models; *Niemelä et al.* (2005) compare static simulation with live measurements (see below); *Laiho et al.* (2002) observe that static and dynamic simulation agree on the mean values of, for example, transmit powers, but the distributions are different. We cite some investigations on fine-tuning of detailed parameters below (more are referenced by *Laiho et al.*, 2002). We will use static simulation as the reference method for performance evaluation.

2.4 The classical static model

We formally introduce an adapted version of the classical static model as described, for example, by *Hoshyar et al.* (2003) and by *Eisenblätter et al.* (2002). We largely adopt the notation of *Eisenblätter et al.*

Let us introduce some basic notation first. The network consists of a fixed set $\mathcal{N}$ of cells installed in some planning area A. Usually, two-dimensional areas $A \subset \mathbb{R}^2$ are considered, but three-dimensional areas $A \subset \mathbb{R}^3$ are also possible. We denote the size of the network (the number of cells) by $n := |\mathcal{N}|$. Within a snapshot, a fixed set M of users wishes to access service, and it is assumed that no user finishes a call or initiates a new one within the observation time. Each user is requesting service by some fixed cell $i \in \mathcal{N}$; the set of all users requesting service from Node-B i is denoted by M_i. Although it is possible to consider soft handover in static models, we do not do so here for the sake of simplicity. Under these auspices, $\{M_i\}_{i \in \mathcal{N}}$ is a partition of M (*i.e.*, each element of M is contained in some set M_i, and the sets M_i are disjoint). The association of users to cells and potential sources of strong interference are illustrated in Fig. 2.6.

2.4.1 CIR equations for dedicated channels

The core component of the model is a linear equation system governing transmit powers derived under the assumption of *perfect power control.*

Perfect power control. To represent the power control mechanism by a single constraint, the assumption of *perfect power control* is made. Within the observation time of a snapshot, we assume the right-hand side threshold of (2.1) to be fixed and not adjusted by the outer loop of power control. The threshold is called the user's CIR *target*, we denote it by $\mu_m^{\uparrow} \geq 0$ for uplink and $\mu_m^{\downarrow} \geq 0$ for downlink. The CIR target corresponds precisely to the desired E_b/N_0 target for the user adjusted by the applicable processing gain. A mobile maintaining a dedicated link in only one direction has a CIR target of 0 in the other direction. (Reverse-link signaling communication is usually ignored.) The inner loop of power control is assumed to perfectly equalize all fluctuations of the signal caused by fast fading. The graph in Fig. 2.5(b) is thus approximated by a horizontal line and the inequality "$\geq$" in (2.1) is met with equality.

Dynamic link-level simulation provides sensible values for the CIR target (e. g., *Neto et al.*, 2002; *Olmos & Ruiz*, 2000; *Optimi, Inc*, 2006*b*); it only considers a single transmission link, as opposed to a number of cells and concurrent transmissions. *Sippilä et al.* (1999) investigate the behavior of imperfect power control and show how CIR targets can be calibrated to achieve more accurate results in static models. *Leibnitz* (2003) models the power control loop analytically and investigates its impact on system performance.

Channel gain. The channel gain between a Node-B and the mobile (including intermediate hardware) is subsumed in a factor, which is specific to a pair of sender and receiver and the link direction; it is assumed constant during the observation time. Channel gain is divided into a path loss component and a shadow fading component, see Sec. 2.1.2. We assume that both components depend on the cell $i \in \mathcal{N}$ and the user's location. Two users at precisely the same location then have the same channel gain towards all cells. The potential channel gain values to and from a cell $i \in \mathcal{N}$ are specified in the functions $\gamma_i^{\uparrow}, \gamma_i^{\downarrow} : A \to [0, 1]$ defined over the planning area. The multiplicative shadowing component $\tilde{\gamma}_i : A \to \mathbb{R}_{>0}$ is assumed to be identical for uplink and downlink and only depend on user position. (This is only to simplify notation, see the discussion on p. 30.) The complete channel gain for point $x \in A$ is $\gamma_{ix}^{\uparrow} := \gamma_i^{\uparrow}(x)\tilde{\gamma}(x)$ in uplink and $\gamma_{ix}^{\downarrow} := \gamma_i^{\downarrow}(x)\tilde{\gamma}(x)$ in downlink. For a user m located at $x \in A$, we use the shorthands $\gamma_{im}^{\uparrow} := \gamma_{ix}^{\uparrow}$ and $\gamma_{im}^{\downarrow} := \gamma_{ix}^{\downarrow}$.

The impact of different prediction methods on the accuracy of static simulations is investigated by comparison with live measurements by *Niemelä et al.* (2005); they conclude that the COST 231-Hata model tends to overestimate capacity, while ray-tracing models underestimate it.

Activity factors. It is a feature of UMTS, that no empty data packets are transmitted; if there is no information to be sent, the channel remains idle. User activity therefore influences resource utilization. *Brady* (1969) first modeled user activity in voice conversations; his result is that the active periods are well modeled by exponentially distributed state durations. *Stern et al.* (1996) refined the model for the use with new-generation communication systems. *Tran-Gia et al.* (2001) give an overview of source traffic models for complex data applications.

In static models, short-term user activity patterns are represented by scaling the interference created by a link with the *activity factor,* which specifies the fraction of time that a connection is active during the observation time. It is denoted in the following by the parameter $\alpha_m^{\uparrow}$ and $\alpha_m^{\downarrow}$ for user m in uplink and downlink. The two parameters may be distinct, they are specific to a user connection and take on values in the interval $[0,1]$. The average activity approach tends to overestimate cell capacity (*Nordling,* 2005); the model is, however, universally used by scientists as well as in commercial network planning and optimization software (e.g., *AWE Communications GmbH,* 2005a; *Cosiro GmbH,* 2006).

Uplink CIR equation. We denote the transmission power of mobile m by $p_m^{\uparrow}$. The received signal at cell i is hence $\gamma_{mi}^{\uparrow}p_m^{\uparrow}$—in periods of (uplink) activity of mobile m. The *average* power received at cell i from a mobile m over the observation time is calculated as $\gamma_{mi}^{\uparrow}\,\alpha_m^{\uparrow}\,p_m^{\uparrow}$. The average carrier-to-interference ratio of a link from mobile m to cell i is the ratio of the received signal (in active an period) over the sum of the average powers received by cell i from the remaining users $k \in M$, $k \neq m$ and noise, which we denote by $\eta_i^{\uparrow}$. The uplink CIR is thus

$$\frac{\gamma_{mi}^{\uparrow}p_m^{\uparrow}}{\sum_{k\neq m}\gamma_{ki}^{\uparrow}\,\alpha_k^{\uparrow}\,p_k^{\uparrow}+\eta_i^{\uparrow}}\,. \tag{UL CIR}$$

For shorter notation, we introduce the *average total received power* $\bar{p}_i^{\uparrow}$ at cell i comprising the average power received from all mobiles in the network plus noise:

$$\bar{p}_i^{\uparrow} := \sum_{m\in M}\gamma_{mi}^{\uparrow}\,\alpha_m^{\uparrow}\,p_m^{\uparrow}+\eta_i^{\uparrow}\,. \tag{2.2}$$

The fundamental assumption of perfect power control is that during active periods of a connection the inner power control loop adjusts $p_m^{\uparrow}$ such that the expression (UL CIR) matches the CIR target $\mu_m^{\uparrow}$. The following CIR equality holds:

$$\frac{\gamma_{mi}^{\uparrow}\,p_m^{\uparrow}}{\bar{p}_i^{\uparrow}-\gamma_{mi}^{\uparrow}\,\alpha_m^{\uparrow}\,p_m^{\uparrow}}=\mu_m^{\uparrow}\,. \tag{2.3}$$

Note that the expression on the left-hand side is equivalent to (UL CIR).

Orthogonality. As discussed on pp. 13–14, all downlink transmissions stemming from the same cell are encoded using orthogonal codes (which we assume to be available in sufficient quantity), but under realistic channel conditions orthogonality is partly lost. This is taken into account by adding the own cell's interference to the denominator in CIR calculations adjusted by an *orthogonality loss factor*[6] $\omega \in [0,1]$. A value

[6]Note that the classical orthogonality factor is the complement $1-\omega$ of the orthogonality loss factor.

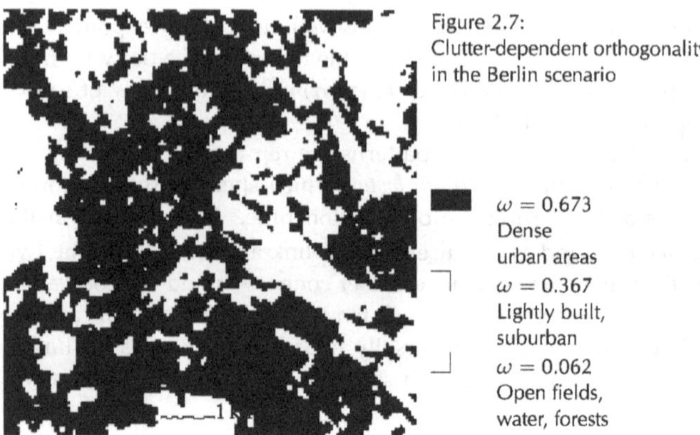

Figure 2.7:
Clutter-dependent orthogonality in the Berlin scenario

of $\omega = 0$ means perfect orthogonality, whereas $\omega = 1$ means orthogonality is lost altogether. This corresponds to assuming a higher processing gain against signals interfering from the own cell.

In simple data models, a uniform value is used for ω. Sophisticated models use a channel specific orthogonality factor. The channel conditions are either modeled specific to a mobile and its environment or to a pair of sender and receiver (*Burger et al.*, 2005). In the data used for our computations, the factor depends on the mobile's location only; we therefore denote the factor applicable to the connection of a cell to a mobile m by ω_m. The distribution of the orthogonality-loss factor for the Berlin scenario is illustrated in Fig. 2.7. Orthogonality is less preserved in densely built areas and more so in environments with no significant multipath propagation (for example microcells, *Hedberg & Parkvall*, 2000). *Hunukumbure et al.* (2002), *Passerini et al.* (2001), and *Pedersen & Mogensen* (2002) investigate orthogonality modeling for static models in more detail.

Downlink CIR equation. We denote the powers emitted on dedicated links by $p^{\downarrow}_{im}$ for a link from cell i to mobile m. In addition to the dedicated channels, the cell transmits on the common channels. We denote the total transmit power emitted by cell i on common channels by $p^{(c)}_i$. (This value includes also the pilot power $p^{(p)}_i$ introduced below.) The average total power emitted by cell i in downlink during the observation time is

$$\bar{p}^{\downarrow}_i := \textstyle\sum_{m \in M_i} \alpha^{\downarrow}_m \, p^{\downarrow}_{im} + p^{(c)}_i \,. \tag{2.4}$$

Using the orthogonality-loss factor, the average total interference power received from signals from the own cell applicable to the connection between i and m is

$$\gamma^{\downarrow}_{im} \, \omega_m \left(\bar{p}^{\downarrow}_i - \alpha^{\downarrow}_m \, p^{\downarrow}_{im} \right) .$$

Orthogonality only applies to signals in the own cell. The interference from all remaining transmitters is calculated as $\sum_{j \neq i} \gamma^{\downarrow}_{jm} \, \bar{p}^{\downarrow}_j$. In addition, some noise specific to user equipment and location occurs, which we denote by $\eta_m^{\downarrow}$. The complete CIR equation for the downlink then is

$$\frac{\gamma_{im}^{\downarrow}\, p_{im}^{\downarrow}}{\gamma_{im}^{\downarrow}\, \omega_m\, (\bar{p}_i^{\downarrow} - \alpha_m^{\downarrow}\, p_{im}^{\downarrow}) + \sum_{j \neq i} \gamma_{jm}^{\downarrow}\, \bar{p}_j^{\downarrow} + \eta_m^{\downarrow}} = \mu_m^{\downarrow}\,. \qquad \text{(DL CIR)}$$

2.4.2 Technical constraints

For the CIR-equalities to be meaningful, additional constraints have to be fulfilled that guarantee that the dedicated links can actually be established. There are two kinds of constraints: power limits reflecting the capabilities of the equipment and coverage constraints ensuring proper decoding of the pilot channel at each mobile.

Uplink coverage. A mobile handset has a limited transmit power, we denote the maximum possible value by by $p_{\max}^{(\text{UE})} > 0$. This introduces the constraint $p_m^{\uparrow} \leq p_{\max}^{(\text{UE})}$ for all $m \in M$. For clarity of notation, we assume that the value is identical for all mobiles. In reality, $p_{\max}^{(\text{UE})}$ differs with the type of the mobile. The UMTS standard specification (3GPP TSG RAN, 2006*b*) defines four user equipment power classes. The weakest class, power class 4, has a maximum output power of $p_{\max}^{(\text{UE})} = 21\,\text{dBm}$; most handsets conform to power class 4.

The constraint on the mobile's transmit power introduces a dependency of the cell range on the received cell power; this phenomenon is called *cell breathing*. The higher the received cell power is, the higher is the contribution of the mobile that is needed to ensure that equality holds in (2.3). The maximum attenuation that the mobile can overcome is therefore smaller if the cell is more loaded. In urban areas, this is less of an issue than in rural areas, where cells tend to be larger. The effect is illustrated in Fig. 2.8.[7] In addition to the cell's current received power, the minimum attenuation value depends on the user's CIR-target, so the cell's maximum area differs according to the desired service. Cell breathing also occurs in downlink, but it is less problematic because base stations have a higher maximum output power than mobiles.

Cell power limits and soft capacity. A cell's transmission unit has a maximum output power value, which we denote by $p_{\max}^{\downarrow}$. A typical value for a macro cell is $p_{\max}^{\downarrow} = 20\,\text{W}$. However, the constraint on the cell's average downlink transmit power introduced into a static model is more restrictive. The limiting value for the average is a fraction of typically 50–90 % of $p_{\max}^{\downarrow}$ (*Lustig et al.*, 2004); we denote it by $\bar{p}_{\max}^{\downarrow}$ and fix the parameter at 70 %. The gap is required for situations in which more than the average output power is temporarily needed to serve the present users, for example, if a fade needs to be made up for by power control.

The uplink *noise rise* $\text{NR}_i^{\uparrow}$ at an uplink receiver i is the ratio of the total (average) received power to noise:

$$\text{NR}_i^{\uparrow} := \bar{p}_i^{\uparrow} / \eta_i^{\uparrow}\,. \qquad (2.5)$$

[7]Central Lisbon, max. noise rise 5 dB, service with $\mu^{\uparrow} = -14.5\,\text{dB}$, $\eta^{\uparrow} = -100.17\,\text{dBm}$, $p_{\max}^{(\text{UE})} = 21\,\text{dBm}$.

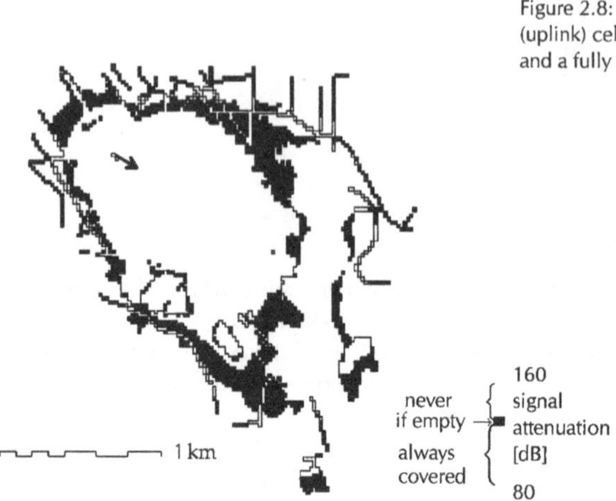

Figure 2.8: Cell breathing—the maximum (uplink) cell size differs for an empty and a fully loaded cell

It is bounded for three reasons: First, the stability of the power control mechanism is at risk if the working point is far away from the base noise. Second, any increase in the received power reduces uplink coverage due to cell breathing. Third, the Node-B's receiver can only function properly up to a certain level of received power. The maximum noise rise translates into a constraint on $\bar{p}_i^{\uparrow}$; we use the symbol $\bar{p}_{\max}^{\uparrow}$ to denote the limiting value. We use a maximum noise rise of 3 dB, which (roughly) corresponds to $\bar{p}_{\max}^{\uparrow} = 2\eta^{\uparrow}$; in practice, values in the range of 3–6 dB are typical (*Lustig et al.*, 2004). Individual limits per cell may also be used as a means for achieving better coverage through load balancing (*Hedberg & Parkvall*, 2000).

The power constraints in uplink and downlink introduce *soft capacity* constraints on a cell, as the resources spent on serving a given user vary with the level of interference received from other cells. There are additional *hard* capacity limitations such as processing power at the Node-B or code availability; we omit these.

Pilot channel coverage: E_c and E_c/I_0. As discussed in Sec. 2.2.2, a mobile needs to receive a pilot signal of sufficient strength E_c and quality E_c/I_0 for having coverage. Let $p_i^{(P)}$ denote the strength of the pilot signal emitted by cell i; it is not dynamically adjusted but a parameter fixed by the operator. We denote the threshold by $\pi_{E_c} \geq 0$; the condition for E_c coverage of a mobile m is:

$$\gamma_{im}^{\downarrow} p_i^{(P)} \geq \pi_{E_c} .$$

An example plot of the E_c value and the E_c-covered area (for a threshold of $\pi_{E_c} = -85$ dBm) in a part of the Lisbon scenario is shown in Fig. 2.9(a). The condition for E_c/I_0 coverage is another constraint of type (2.1). The E_c/I_0 value is calculated as the ratio of the received signal strength over the sum of (downlink) noise $\eta_m^{\downarrow}$ at the receiver m and the total of all received average powers. We denote the threshold value

Figure 2.9: Coverage in a UMTS network

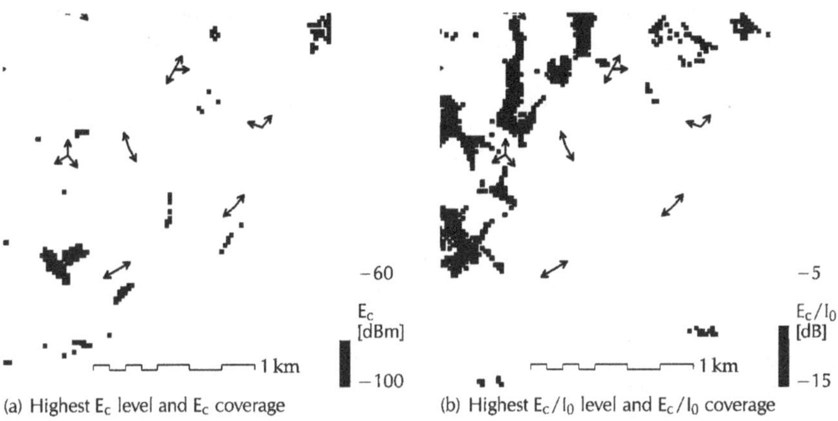

(a) Highest E_c level and E_c coverage (b) Highest E_c/I_0 level and E_c/I_0 coverage

that has to be met for sufficient signal quality by μ_{E_c/I_0}; the full condition is

$$\frac{\gamma^{\downarrow}_{im} p^{(P)}_{i}}{\eta^{\downarrow}_{m} + \sum_{j \in \mathcal{N}} \gamma^{\downarrow}_{jm} \bar{p}^{\downarrow}_{j}} \geq \mu_{E_c/I_0} .$$

Note that the pilot channel is not power controlled, so we have a genuine inequality. A plot of the pilot signal quality E_c/I_0 and the E_c/I_0-covered area in a part of the Lisbon scenario is sketched in Fig. 2.9(b) (for $\mu_{E_c/I_0} = -10\,\text{dB}$). The two areas are not identical in general; coverage is only provided in their intersection.

2.4.3 Formal system model

We summarize the static model of a UMTS radio access network for easier reference and formalize some notions.

Vector notation. We adopt the following notational conventions: if a scalar x_i is defined for all indices i in an index set S, we write $\boldsymbol{x}$ for the (column) vector of matching dimension consisting of the elements $(x_i)_{i \in S}$ (usually $\boldsymbol{x} \in \mathbb{R}^{|S|}_{+}$). For scalars having two indices y_{ij} from the same index set S, we shall write $\boldsymbol{Y}$ for the square matrix $(y_{ij})_{i,j \in S}$ consisting of these elements. Vector inequalities like $\boldsymbol{a} \geq \boldsymbol{b}$ are to be understood component-wise, *i. e.*,

$$\boldsymbol{a} \geq \boldsymbol{b} :\Leftrightarrow a_i \geq b_i \quad \text{for all } i \in S .$$

For two vectors $\boldsymbol{a}, \boldsymbol{b} \in \mathbb{R}^{|S|}$, we define the vector interval $[\boldsymbol{a}, \boldsymbol{b}]$ as the set of vectors $\boldsymbol{x}$ fulfilling both $\boldsymbol{a} \leq \boldsymbol{x}$ and $\boldsymbol{x} \leq \boldsymbol{b}$. The half open intervals $(\boldsymbol{a}, \boldsymbol{b}]$, $[\boldsymbol{a}, \boldsymbol{b})$ and the open interval $(\boldsymbol{a}, \boldsymbol{b})$ are defined analogously but with strong inequalities at the respective sides. We write $\boldsymbol{x}^t$ for the transposed (row) vector of $\boldsymbol{x}$. We will occasionally use the plain form x for a single representative but unspecified element of the vector $\boldsymbol{x}$. We will also write $\mathbb{R}_+ := \{x \in \mathbb{R} \mid x \geq 0\}$ and $\mathbb{R}_{>0} := \{x \in \mathbb{R} \mid x > 0\}$. The symbols $\mathbf{1}$ and $\mathbf{0}$ denote column vectors or matrices of matching dimension consisting of all ones and all zeroes, respectively. The symbol $\boldsymbol{I}$ denotes the identity matrix of matching (square) dimension.

We first define a radio network and a snapshot independently; the concepts are then linked through the definition of E_c coverage and a feasibility system describing whether a network can serve a snapshot.

Definition 1: A *radio network* formed by a set of cells $\mathcal{N}$, with $|\mathcal{N}| =: n$ is described by measurable uplink and downlink path loss functions $\gamma_i^{\uparrow}, \gamma_i^{\downarrow} : A \to [0,1]$ for all $i \in \mathcal{N}$ and the following vectors of dimension n: maximum transmit powers $\bar{\boldsymbol{p}}^{\downarrow}_{\text{max}} > \mathbf{0}$; pilot channel power $\boldsymbol{p}^{(\text{P})} \in (\mathbf{0}, \bar{\boldsymbol{p}}^{\downarrow}_{\text{max}}]$; total common channel power $\boldsymbol{p}^{(\text{C})} \in [\boldsymbol{p}^{(\text{P})}, \bar{\boldsymbol{p}}^{\downarrow}_{\text{max}}]$; noise levels at the cells $\boldsymbol{\eta}^{\uparrow} > \mathbf{0}$; and maximum receive powers $\bar{\boldsymbol{p}}^{\uparrow}_{\text{max}} \in (\boldsymbol{\eta}^{\uparrow}, \infty)$.

Definition 2: A *snapshot* of a set M of mobiles is described by a mapping of each mobile to a location in the area A, the shadow-fading function $\tilde{\gamma} : A \to \mathbb{R}_+$ and the following parameter vectors of dimension $|M|$: noise levels at the receiver $\boldsymbol{\eta}^{\downarrow} \geq \mathbf{0}$, maximum transmit powers $\boldsymbol{p}^{(\text{UE})}_{\text{max}} \geq \mathbf{0}$, CIR targets in uplink and downlink $\boldsymbol{\mu}^{\uparrow}, \boldsymbol{\mu}^{\downarrow} \geq \mathbf{0}$, activity factors in uplink and downlink $\boldsymbol{\alpha}^{\uparrow}, \boldsymbol{\alpha}^{\downarrow} \in (\mathbf{0}, \mathbf{1}]$, and orthogonality-loss factors $\boldsymbol{\omega} \in [\mathbf{0}, \mathbf{1}]$.

According to the model of shadow fading, the function $\tilde{\gamma}$ may depend on the network configuration; we omit this detail for the sake of brevity. We deliberately exclude $\alpha^{\downarrow}_m, \alpha^{\uparrow}_m = 0$ because these cases would necessitate technical case distinctions in the upcoming lemmas, but this is no restriction. If a user should not appear in one link direction, setting the CIR target to zero has basically the same effect as a vanishing activity factor. The requirement that $\gamma_i^{\uparrow}, \gamma_i^{\downarrow}$ be measurable (for a definition, see *Klenke*, 2006) is needed in Lemma 4.1.

Definition 3: For any network and any snapshot, a partition $\{M_i\}_{i \in \mathcal{N}}$ of the set of M represents an *E_c-covered* assignment of users to cells, if

$$\begin{aligned} \gamma^{\downarrow}_{im} p^{(\text{P})}_i &\geq \pi_{\text{E}_c} > 0 && \text{for all } i \in \mathcal{N},\ m \in M_i\,, \\ \gamma^{\uparrow}_{mi} &> 0 && \text{for all } i \in \mathcal{N},\ m \in M_i\,. \end{aligned} \tag{2.6}$$

We will say that the snapshot is E_c-covered (by the network) if it is equipped with an E_c-covered assignment.

Feasibility systems. A network is able to serve the users in a snapshot in uplink if it is E_c-covered and there exist (uplink) link powers $p^{\uparrow}_m$ for all $m \in M$ and average cell receive powers $\bar{p}^{\uparrow}_i$ for all $i \in \mathcal{N}$ that fulfill the linear feasibility system

$$\gamma^{\uparrow}_{mi}\, p^{\uparrow}_m = \mu^{\uparrow}_m (\bar{p}^{\uparrow}_i - \gamma^{\uparrow}_{mi}\, \alpha^{\uparrow}_m\, p^{\uparrow}_m) \qquad \text{for all } i \in \mathcal{N}, m \in M_i\,, \tag{2.7a}$$

$$\bar{p}^{\uparrow}_i = \textstyle\sum_{m \in M} \gamma^{\uparrow}_{mi}\, \alpha^{\uparrow}_m\, p^{\uparrow}_m + \eta^{\uparrow}_i \qquad \text{for all } i \in \mathcal{N}\,, \tag{2.7b}$$

$$0 \leq p^{\uparrow}_m \leq p^{(\text{UE})}_{\text{max}} \qquad \text{for all } m \in M\,, \tag{2.7c}$$

$$\eta^{\uparrow}_i \leq \bar{p}^{\uparrow}_i \leq \bar{p}^{\uparrow}_{\text{max}} \qquad \text{for all } i \in \mathcal{N}\,. \tag{2.7d}$$

A network is able to serve the users in a snapshot in downlink if it is E_c-covered and there exist (downlink) link powers $p^{\downarrow}_{im}$ for all $m \in M$, $i \in \mathcal{N}$ and average cell transmit powers $\bar{p}^{\downarrow}_i$ for all $i \in \mathcal{N}$ that fulfill the linear feasibility system

$$\gamma^{\downarrow}_{im}\, p^{\downarrow}_{im} = \mu^{\downarrow}_m \big(\gamma^{\downarrow}_{im}\, \omega_m\, (\bar{p}^{\downarrow}_i - \alpha^{\downarrow}_m\, p^{\downarrow}_{im}) + \sum_{j \neq i} \gamma^{\downarrow}_{jm}\, \bar{p}^{\downarrow}_j + \eta^{\downarrow}_m\big) \quad \text{for all } i \in \mathcal{N},\, m \in M_i\,, \tag{2.8a}$$

$$\bar{p}^{\downarrow}_i = \textstyle\sum_{m \in M_i} \alpha^{\downarrow}_m\, p^{\downarrow}_{im} + p^{(c)}_i \quad \text{for all } i \in \mathcal{N}\,, \tag{2.8b}$$

$$\gamma^{\downarrow}_{im} p^{(p)}_i \geq \mu_{E_c/I_0} \big(\eta^{\downarrow}_m + \textstyle\sum_{j \in \mathcal{N}} \gamma^{\downarrow}_{jm} \bar{p}^{\downarrow}_j\big) \quad \text{for all } i \in \mathcal{N},\, m \in M_i\,, \tag{2.8c}$$

$$p^{(c)}_i \leq \bar{p}^{\downarrow}_i \leq \bar{p}^{\downarrow}_{\max} \quad \text{for all } i \in \mathcal{N}\,. \tag{2.8d}$$

We will characterize the existence of a solution of (2.7) and (2.8), respectively, in Sec. 3.1 below, and we will see that there can be at most one solution. For a complete view on the system, the uplink and downlink components need to be considered jointly. Because in network planning one of the two directions can usually be identified as the bottleneck, we will treat them separately from now on.

2.5 Performance evaluation with static simulation

Static simulation, *i.e.*, Monte Carlo simulation based on a static model, is the state of the art for accurate performance evaluation, at least in commercial tools (*Cosiro GmbH*, 2006; *Ericsson AB*, 2006, 2007; *Forsk SA*, 2007; *Lustig et al.*, 2004; *Optimi, Inc*, 2006*b*; *Schema Ltd*, 2003*b*). We shall now outline the general principle and its application to radio network evaluation. Furthermore, we will detail two substeps that are crucial for the next two chapters: drawing a random snapshot and emulating load control for a given snapshot.

2.5.1 Monte Carlo simulation

Monte Carlo simulation is a technique that uses repeated random trials for calculating quantities that cannot be described in closed-form, or for which such a description cannot be solved, in particular, if random processes are involved (*Itô*, 1987). It first occurred to S. M. Ulam in 1946 when trying to assess his chances of a successful solitaire game (*Eckhardt*, 1987). The method's theoretic aspects are treated in the books by *Fishman & Fishman* (1996) and *Madras* (2002). *Türke* (2006) gives a detailed account in the context of radio network performance evaluation.

Fig. 2.10 illustrates the principle: a sample from the underlying random distribution of snapshots is drawn; the performance of the network for the current snapshot is evaluated; the process is repeated until the empirical distribution of the output quantities describing network performance converge. The method yields accurate results if sensible convergence criteria are chosen, but it often requires many iterations.

In the evaluation of a snapshot, some subset of users that can be served is found. This process is outlined in more detail in Sec. 2.5.3. If service is impossible for some users, the individual reason is registered. Potential reasons are *outage*, a violation of the coverage conditions (2.6), (2.7c), or (2.8c), or *blocking*, a violation of the cell

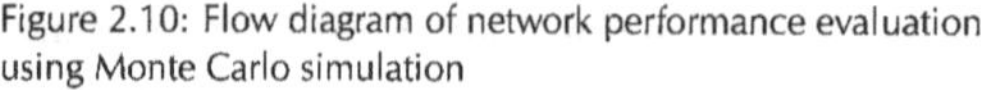

Figure 2.10: Flow diagram of network performance evaluation using Monte Carlo simulation

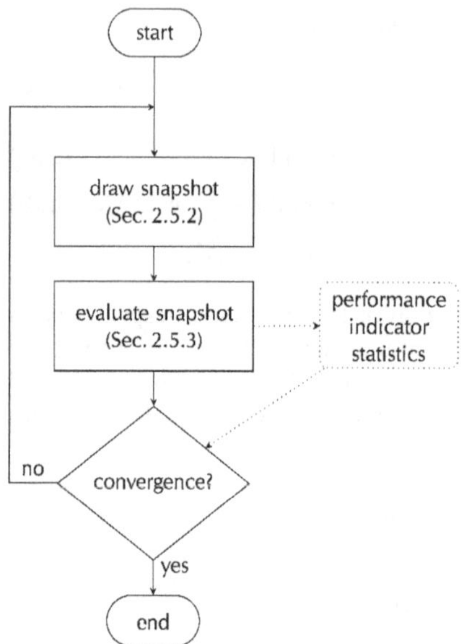

power constraints (2.7d) or (2.8d). For the users that are served, the connection quality, the necessary link powers, and the resulting cell powers are stored. A detailed bookkeeping of all this information produces estimates for the stochastics of network performance in each step. There is no fixed canon of radio network performance indicators; we define ours in Sec. 3.4. We give details on the relevant stochastic aspects and specify our reference implementation of Monte Carlo simulation in Sec. 4.1.

Some aspects of performance fall out of the scope of a static model; this applies to the performance of packet-switched data services, of HSPA, and of dynamic network functions like handover and congestion control. For data services, the fundamental problem is that a snapshot represents an average over a few seconds, while network operation for data users and relevant stochastic effects vary over a time of several minutes; this is, for example, grasped in the web traffic model by the (ETSI, 1997). In consequence, meaningful estimates for average data rate, throughput, or packet delay cannot be obtained with a static model. HSPA, on the other hand, specifically exploits short-term variations of the radio channel (*Holma & Toskala*, 2006), which are eliminated by averaging in a static model. Furthermore, a classical static model is unable to reproduce the dynamic processes that trigger, for example, handover or load control. *Lamers et al.* (2003) attempt to remedy the last aspect with the technique of short-term dynamic simulation featuring sequences of dependent snapshot over time. Performance evaluation for packet-switched services and HSPA requires special tools like, for example, the one by *AWE Communications GmbH* (2005*b*).

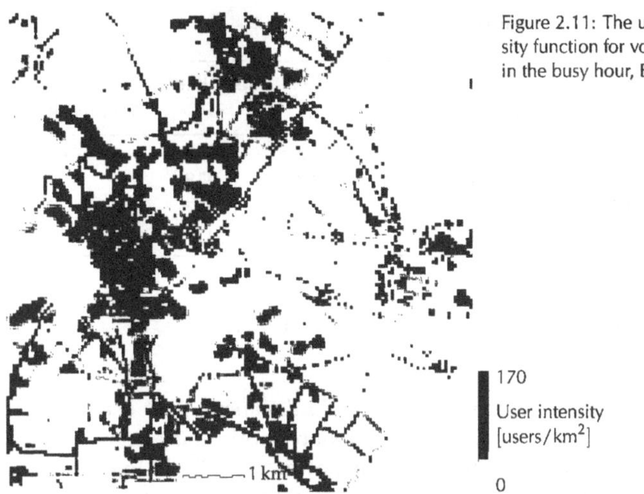

Figure 2.11: The user intensity function for voice service in the busy hour, Berlin

2.5.2 The random distribution on snapshots

We base our description of the random distribution on snapshots on the public data compiled by the MOMENTUM *Project* (2003) as described by *Eisenblätter et al.* (2004). Similar concepts are used in related scientific work (e.g., *Balzanelli et al.*, 2003; *Staehle*, 2004; *Türke & Koonert*, 2005). Three random influences are relevant: the spatial distribution of users, the services they require, and shadow fading. We will discuss these aspects separately and then formalize the standard random model.

Spatial user distribution. User demand varies with the area under consideration and the hour of day. Network operators aim at designing the network to accommodate the user demand in the *busy hour*, which is defined as "the sliding 60-minute period during which occurs the maximum total traffic load in a given 24-hour period" (ATIS *Committee T1A1*, 2001). The 24-hour period is usually a normal business day; extraordinary events are not taken into account.

The number and location of users in a given area define an inhomogeneous spatial *Poisson point process* (*Resnick*, 1992, Ch. 4). This means that there is a *user intensity function* $T : A \to \mathbb{R}_+$ such that for any measurable subset $S \subset A$ the number of users in S is Poisson-distributed with parameter $\int_S T(x)\mathrm{d}x$. Furthermore, the number of users in any (finite) collection of disjoint measurable subsets of A is independent from each other. An example for a realistic user intensity function in the Berlin scenario is shown in Fig. 2.11. A probability variable X taking on values in $\mathbb{N} = \{0, 1, 2, \dots\}$ is Poisson distributed with parameter χ, if the probability of a number m to occur is

$$P(X = m) = \chi^m e^{-\chi} / m!\,.$$

The first and second order moments of a Poisson-distributed random variable are

$$\mathbb{E}(X) = \chi\,, \qquad \mathrm{Var}(X) = \chi\,.$$

Table 2.1: Example services and their parameters

	activity factor $\alpha^{\uparrow}/\alpha^{\downarrow}$	data rate [kbps]	CIR target [dB] $\mu^{\uparrow}$	$\mu^{\downarrow}$	user load $\ell^{\downarrow}$[a]	max users /cell (DL)[b]
Voice	0.5	12.2	−19.5	−17.5	0.009	48
Video	1.0	64.0	−14.6	−12.1	0.058	7
File Download	0.9	64.0	–	−12.6	0.047	9
www	0.9	32.0	–	−15.6	0.024	17

[a] Defined in Sec. 3.1
[b] In a typical urban cell, parameters $\bar{p}^{\downarrow}_{\max} = 14\,\mathrm{W}$, $p^{(c)} = 2\,\mathrm{W}$, $\bar{\tau}^{\downarrow} = 1$ (see (3.37))

Services. A planning scenario contains a set of *service definitions.* Each user requires a service according to a discrete probability distribution, which may depend on the location. For example, in business areas users might be more likely to check their e-mail whereas in residential areas the streaming of television content might be more popular. The service governs the CIR-targets μ and the activity factors α of a user. A sample of services defined in the MOMENTUM data and their parameters is shown in Tab. 2.1. The "Voice" service requires the smallest data rate, which is reflected in the low CIR targets. The uplink CIR $\mu^{\uparrow}$ is lower than the downlink value $\mu^{\downarrow}$ because of the better computing hardware at the Node-B. The activity factor of $\alpha = 0.5$ reflects that a user on average listens as long as he speaks; in a video call, full activity is assumed, as images are transmitted all the time. The "File Download" service has a lower downlink CIR target $\mu^{\downarrow}$ than "Video" at the same data rate because it is not delay sensitive, so decoding errors can be handled by retransmits. Different services result in a varying granularity of users: The last column in Tab. 2.1 shows that a typical cell can serve up to 48 speech users, but only 7 video users simultaneously.

Shadow fading. While fast fading is assumed to be equalized by perfect power control, the shadow fading component $\tilde{\gamma}$ at a given location is commonly modeled as a lognormally distributed random variable (*Rappaport*, 2002; *Saunders*, 1999). The logarithmic mean is usually 0 dB; for its logarithmic standard deviation, a range of 6–12 dB is common. Simple shadowing models use independent values per user; more accurate schemes include a correlation between the different links to the same receiver. The shadowing values are usually identical for users within a small (parameterized) distance of each other. We use the model with a cross-correlation adapted to the angle between the mobile and the different antennas described by *Winter et al.* (2003) with a standard deviation of 8 dB in our notations; for simpler notation we do not express the dependence of the shadowing values on the network formally but consider $\tilde{\gamma}$ to be a function over the planning area.

The random distribution on snapshots assumed in this thesis is the following:

Definition 4: A randomly drawn snapshot follows the *standard random distribution (on snapshots)*, if

(a) Each user is assigned to a finite set R of services, and all users of one service $r \in R$ have identical CIR targets $\mu_r^{\uparrow}, \mu_r^{\downarrow} \geq 0$ and activity factors $\alpha_r^{\uparrow}, \alpha_r^{\downarrow} \in (0, 1]$.
(b) The expected number and position of users of a service $r \in R$ follow an inhomogeneous Poisson point process with intensity function $T_r : A \to \mathbb{R}_+$ and $\int_A T_r(x)\mathrm{d}x < \infty$.
(c) There are measurable functions $\eta^{\downarrow} : A \to \mathbb{R}_{>0}$ and $\omega : A \to [0,1]$ such that a user at position $x \in A$ has a noise level of $\eta^{\downarrow}(x)$ and an orthogonality loss factor of $\omega(x)$.

If, in addition, $\tilde{\gamma} \equiv 1$, *i. e.*, random shadowing is excluded, we say that the snapshot follows the *restricted random distribution (on snapshots)*.

2.5.3 Static load control

Evaluation based on static models can only mimic realistic, dynamic congestion control. Some such mechanism is, however, needed in any network planning software that produces accurate performance predictions. In the context of a static model, we will speak of (static) load control in the following. Practice-relevant scientific contributions are rare in this field; our overview is based on the recent thesis by *Türke* (2006). Some related literature is cited in the beginning of the next chapter.

Static evaluation schemes decide whether or not service is possible for each user and generally aim at maximizing the set of served users. There might be several options for service; for example, a speech connection can use various speech codecs, or a data connection can have various data rates. We do not consider these subtleties but assume that a binary (admit/block) decision is taken per user. Load-control routines generally try to find a "maximal" subset of users for which the constraint system is feasible because the network's resources should be fully used, but there is no standardized notion of maximality; depending on the model, a "maximal" feasible set of users need not exist. We do not delve into these considerations, however, because our upcoming developments render them irrelevant.

An archetypical template for a load-control scheme is sketched in Fig. 2.12. The algorithm starts with the complete set of users and computes the solution to the CIR equation system parts of the system models (2.7) and (2.8). It then checks whether any inequality is violated because a user is in outage or a cell is in overload. If so, some user is removed from the system, and the evaluation is repeated. If all constraints are fulfilled for the current set of users, however, the algorithm is not yet finished—it might be the case that a user that was removed earlier on can now be served because a later removal freed the necessary resources. If this is the case, the user has to be admitted to the system again, which could in turn cause further outage or soft blocking in other cells.

Researchers have conceived variations and enhancements of this basic scheme. For example, simple estimates can often spare the effort of evaluating the entire system every time (*Catrein et al.*, 2005), more than one user can be removed at once (*Türke*, 2006), we may start with the empty set and add users gradually instead of removing them from the complete set (*Dehghan et al.*, 2000). Either way, all known implemen-

Figure 2.12: Load control in classical static evaluation

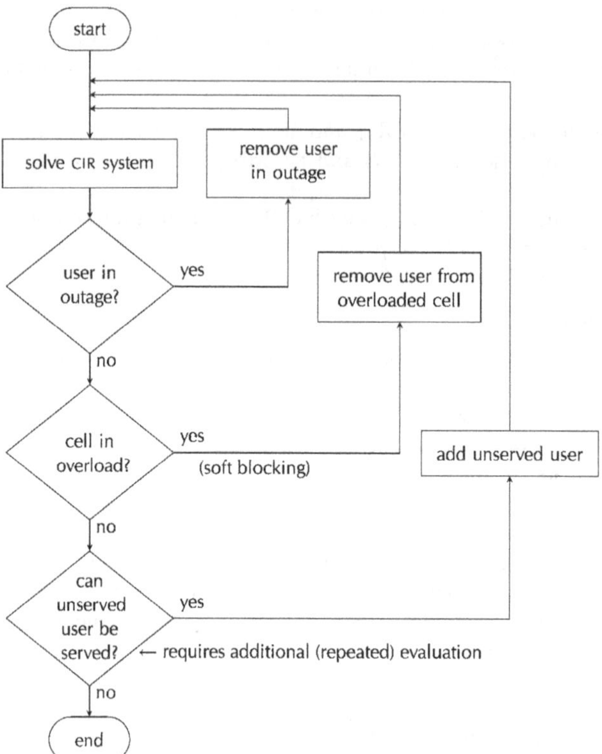

tations of user-specific load-control schemes require repeated evaluations of the CIR equation system.

Things to remember: Classical static performance evaluation

- We consider UMTS W-CDMA radio networks in FDD mode on a single carrier, common channels with fixed power, and traffic on dedicated channels with perfect power control. We neglect soft handover.

- The classical static model in this setting is:

Network parameters			*Snapshot parameters*	
$\mathcal{N}$	set of cells		M/M_i	all mobiles/mobiles in cell i
$p_{\max}$	maximum cell powers		$\eta^{\downarrow}$	noise at mobile
$p^{(\mathrm{P})}, p^{(\mathrm{C})}$	pilot/common channel power		$p^{(\mathrm{UE})}_{\max}$	max mobile transmit power
$\eta^{\uparrow}$	noise at cell		μ	CIR target
$\gamma^{\uparrow}/\gamma^{\downarrow}$	path loss functions		α	activity factor
$\downarrow$			ω	orthogonality-loss factor
$\gamma^{\uparrow}_{mi}/\gamma^{\downarrow}_{im}$	end-to-end channel gain	$\leftarrow$	$\tilde{\gamma}$	shadow fading function

If $\mathrm{E_c}$ coverage holds ($\gamma^{\downarrow}_{im} p^{(\mathrm{P})}_i \geq \pi_{\mathrm{E_c}}$ for all $i \in \mathcal{N},\ m \in M_i$), the network can serve the snapshot in *uplink* if cell powers $(\bar{p}^{\uparrow}_i)_{i\in\mathcal{N}}$ and link powers $(p^{\uparrow}_m)_{m\in M}$ exist that fulfill:

$$
\begin{aligned}
\gamma^{\uparrow}_{mi}\, p^{\uparrow}_m &= \mu^{\uparrow}_m (\bar{p}^{\uparrow}_i - \gamma^{\uparrow}_{mi}\, \alpha^{\uparrow}_m\, p^{\uparrow}_m) && \text{for all } i \in \mathcal{N}, m \in M_i && \text{(CIR UL)}\\
\bar{p}^{\uparrow}_i &= \textstyle\sum_{m\in M} \gamma^{\uparrow}_{mi}\, \alpha^{\uparrow}_m\, p^{\uparrow}_m + \eta^{\uparrow}_i && \text{for all } i \in \mathcal{N} && \text{(cell RX)}\\
p^{\uparrow}_m &\leq p^{(\mathrm{UE})}_{\max} && \text{for all } m \in M && \text{(UL cov)}\\
\eta^{\uparrow}_i \leq \bar{p}^{\uparrow}_i &\leq \bar{p}^{\uparrow}_{\max} && \text{for all } i \in \mathcal{N} && \text{(UL cell cap.)}
\end{aligned}
$$

The network can serve the snapshot in *downlink* can be if cell powers $(\bar{p}^{\downarrow}_i)_{i\in\mathcal{N}}$ and link powers $(p^{\downarrow}_{im})_{m\in M, i\in\mathcal{N}}$ exist that fulfill:

$$
\begin{aligned}
\gamma^{\downarrow}_{im}\, p^{\downarrow}_{im} &= \mu^{\downarrow}_m \Big(\gamma^{\downarrow}_{im}\, \omega_m\, (\bar{p}^{\downarrow}_i - \alpha^{\downarrow}_m\, p^{\downarrow}_{im}) && \text{for all } i \in \mathcal{N}, m \in M_i && \text{(CIR DL)}\\
&\qquad + \sum_{j\neq i} \gamma^{\downarrow}_{jm}\, \bar{p}^{\downarrow}_j + \eta^{\downarrow}_m\Big) && && \\
\bar{p}^{\downarrow}_i &= \textstyle\sum_{m\in M_i} \alpha^{\downarrow}_m\, p^{\downarrow}_{im} + p^{(\mathrm{C})}_i && \text{for all } i \in \mathcal{N} && \text{(cell TX)}\\
\gamma^{\downarrow}_{im} p^{(\mathrm{P})}_i &\geq \mu_{\mathrm{E_c/I_0}} (\eta^{\downarrow}_m + \textstyle\sum_{j\in\mathcal{N}} \gamma^{\downarrow}_{jm} \bar{p}^{\downarrow}_j) && \text{for all } i \in \mathcal{N},\ m \in M_i && (\mathrm{E_c/I_0}\ \text{cov})\\
p^{(\mathrm{C})}_i \leq \bar{p}^{\downarrow}_i &\leq \bar{p}^{\downarrow}_{\max} && \text{for all } i \in \mathcal{N} && \text{(DL cell cap.)}
\end{aligned}
$$

- In the standard random distribution on snapshots, CIR target and activity factor are service-dependent, user positions follow a Poisson point process, and noise and orthogonality depend on user position.

- Expected network performance is traditionally evaluated with Monte Carlo simulation. Load control is implemented as an iterative trial-and-error procedure.

Interference-coupling complementarity systems

We present a concise, closed-form system model that allows network performance evaluation for a given snapshot in short time. The novel *interference-coupling complementarity systems* are of compact dimension and have a built-in notion of soft capacity and load control. With the new models and methods, we can evaluate all relevant indicators of network performance for a given snapshot at low computational cost.

The new system model generalizes the recent dimension reduction technique, which addresses difficulties in solving the CIR equation systems. The equation systems are potentially large, as their size depends on the number of users in the snapshot. In addition, they tend to be ill-conditioned due to the high dynamic range of the input data. For instance, *Amaldi et al.* (2006) report this to be a problem in optimization. Dimension reduction eliminates link power variables and treats users implicitly. The cell powers are described as solutions to *interference-coupling equation systems* having the dimension of the number of cells. This formulation has since become the standard model for UMTS radio network analysis in the research community.

Besides power control, also load control is viewed on cell level in our new model, and users can be treated implicitly throughout. This view is enabled by the novel assumption of *perfect load control*, which complements the usual perfect power control. Formally, we introduce a new set of cell variables into the equation system. The new variables represent the cells' user admission policy and are tied to the cell powers with complementarity conditions. The individual reaction of the cells to overload is thus integrated into the compact system model, and the model describes the coupling relationships among load-controlled and power-controlled cells precisely.

A single evaluation of the new interference-coupling complementarity systems can substitute a lengthy iterative load control scheme if some simplifications are accepted.

The complementarity systems can be solved efficiently by iterative solution algorithms presented below; their computational effort compares to solving the linear coupling system. Individual users are not explicitly considered, however, so the compact load control representation cannot reflect load control decisions per user. Furthermore, outage due to a lack of E_c/I_0 or uplink coverage has to be neglected. Computations in the next chapter will show that the inaccuracies are tolerable in practical settings.

The chapter is structured as follows: In Sec. 3.1, we derive the (well-known) linear interference-coupling systems and show how they relate to the original system model as stated in Sec. 2.4.3. We then discuss the notion of perfect load control, the resulting system model, and our methods for solving them in Sec. 3.2. An alternative representation of the system along the lines of the known pole equations for the capacity of a UMTS cell is derived in Sec. 3.3. In Sec. 3.4, we formally specify the performance indicators relevant for radio network planning in terms of our new model.

Related work. *Hanly* (1999) seems to be the first to describe a dimension reduced equation system of cell powers with no explicit user powers. His model has been adapted and generalized by *Mendo & Hernando* (2001); we rely on their version. It has been subsequently used by most authors for analysis and simulation, for example in (*Lamers et al.*, 2003; *Mäder & Staehle*, 2004; *Nawrocki et al.*, 2005; *Türke*, 2006). The compact systems are hardly used in mathematical optimization methods for network planning; only *Amaldi et al.* (2003) employ a rudimentary version of coupling equation systems for faster computation of cell powers. Beyond transmit power calculation, *Nasreddine et al.* (2006) propose using the derivatives of the coupling matrix for frequency assignment in a multi-carrier W-CDMA-system.

Limits on power levels per cell usually apply, so the coupling equation and the capacity region of a network have been studied. Some results developed in the theory of positive matrices apply; a comprehensive monograph on this topic is the book by *Berman & Plemmons* (1994). *Fischer* (2006) compares methods for estimating the Perron root in similar systems occurring in coupling equations for wireless networks. *Catrein et al.* (2004) derive conditions for the feasibility of serving a given set of users. Power control algorithms to confine network operation to the feasible region are treated in the book by *Stańczak et al.* (2006). *Feiten & Mathar* (2005) use a projection method to this end and also pose the power control problem in a game-theoretic setting. *Feiten* (2006) furthermore develops the idea of scaling user demands (with a network-wide uniform factor) if not all demands can be fulfilled. *Zdunek & Nawrocki* (2006) determine feasible (nonnegative) cell powers in overload situations by solving a least square problem. To the best of our knowledge, no contribution so far develops any notion of how much traffic individual cells sacrifice in order to attain feasible cell powers.

Different methods have been chosen to deal with power limits in radio network evaluation and planning. Some authors assume all cells transmitting at the maximum power; capacity evaluation then aims at determining how many users can fit into the cells under this condition (e. g., *Balzanelli et al.*, 2003; *Siomina & Yuan*, 2004, 2005). Alternatively, any capacity left after serving users on dedicated channels is assumed to be spent on common channels (e. g., *Hoßfeld et al.*, 2006). In some cases, transmit power limits are not imposed (e. g., *Staehle*, 2004), but evaluation is limited to snapshots that can be served with finite powers. Some optimization models for network design decide which users to serve in a mixed integer programming model (*Amaldi et al.*, 2006; *Eisenblätter et al.*, 2002; *Schmeink*, 2005). In downlink evaluation, the idea of solving the CIR equation system iteratively and clipping the transmit powers at the limiting value suggests itself and has been the starting point for developing our model. For example, *Türke* (2006) also uses this approach.

3.1 Linear interference-coupling equation systems

The interference-coupling equation systems are derived for uplink and downlink by eliminating the link-power variables from the CIR equation systems (2.7a)/(2.7b) and (2.8a)/(2.8b), respectively. In the following, we use a classical result, which characterizes the solvability of certain equation systems through the spectral radius (the largest modulus of an eigenvalue) of what we will call the interference-coupling matrix. For example, the book by *Stańczak et al.* (2006, Theorems A.11, A.35) contains a proof.

Lemma 3.1: *For any nonnegative square matrix C and any positive vector v of matching dimension, the equation system*

$$\bar{p} = C\bar{p} + v$$

admits a nonnegative solution $\bar{p}$ if and only if the spectral radius of C is strictly smaller than one. In this case, the series $\sum_{k=0}^{\infty}(C)^k$ converges to the inverse of $(I - C)$, and the solution $\bar{p} = (I - C)^{-1}v$ is unique and strictly positive. □

The series in the theorem is called the Neumann series. Its convergence to the inverse can be seen from

$$(I - C)(I + C + C^2 + C^3 + \cdots) = I - C + C - C^2 + C^2 - C^3 + \cdots,$$

which converges to the unit matrix if the powers of the matrix converge to zero. The spectral radius of a nonnegative matrix is also called its *Perron root*. We will denote the spectral radius of a matrix C by $\rho(C)$.

Uplink coupling equation system. Suppose that for a given network and an E_c-covered snapshot we have a solution of the uplink system model (2.7). The average power received at some antenna i is calculated as

$$\bar{p}_i^{\uparrow} = \sum_{m \in M} \gamma_{mi}^{\uparrow}\, \alpha_m^{\uparrow}\, p_m^{\uparrow} + \eta_i^{\uparrow}, \tag{3.1}$$

and for any $m \in M_j$, the respective CIR equation (2.7a) holds:

$$\gamma_{mj}^{\uparrow}\, p_m^{\uparrow} = \mu_m^{\uparrow}(\bar{p}_j^{\uparrow} - \gamma_{mj}^{\uparrow}\, \alpha_m^{\uparrow}\, p_m^{\uparrow}).$$

The cell indices i and j are different if m is not served by cell i. When solving the CIR equation for the term $\gamma_{mj}^{\uparrow}\, \alpha_m^{\uparrow}\, p_m^{\uparrow}$, we obtain

$$\gamma_{mj}^{\uparrow}\, \alpha_m^{\uparrow}\, p_m^{\uparrow} = \frac{\alpha_m^{\uparrow}\, \mu_m^{\uparrow}}{1 + \alpha_m^{\uparrow}\, \mu_m^{\uparrow}}\, \bar{p}_j^{\uparrow}.$$

This step is valid as long as $\alpha_m^{\uparrow} > 0$, which Def. 2 on p. 26 requires. The strength of the signal emitted by user m will thus always contribute a constant fraction of the average total power $\bar{p}_j^{\uparrow}$ emitted by the serving cell. The ratio depends only on the properties of the user. In analogy to, e.g., *Holma & Toskala* (2001, Ch. 8), we call this ratio the uplink *user loading factor*

$$\ell_m^{\uparrow} := \frac{\alpha_m^{\uparrow} \mu_m^{\uparrow}}{1 + \alpha_m^{\uparrow} \mu_m^{\uparrow}} \geq 0 . \tag{3.2}$$

Typical values for the user load factor of different services are listed in Tab. 2.1 (for downlink). The term related to m in the sum in (3.1) is thus:

$$\gamma_{mi}^{\uparrow} \alpha_m^{\uparrow} p_m^{\uparrow} = \frac{\gamma_{mi}^{\uparrow}}{\gamma_{mj}^{\uparrow}} \gamma_{mj}^{\uparrow} \alpha_m^{\uparrow} p_m^{\uparrow} = \frac{\gamma_{mi}^{\uparrow}}{\gamma_{mj}^{\uparrow}} \ell_m^{\uparrow} \bar{p}_j^{\uparrow} .$$

Note that we required $\gamma_{mj}^{\uparrow} > 0$ for an E_c covered snapshot. For $i = j$, *i. e.*, for the contributions of the mobiles served by cell i itself, the right hand side is simply $\ell_m^{\uparrow} \bar{p}_j^{\uparrow}$.

We now substitute this relation into (3.1). To group the terms in the sum that relate to mobiles being served by the same cell, we define the uplink *interference-coupling coefficients* $(c_{ij}^{\uparrow})_{i,j \in \mathcal{N}}$ for $i, j \in \mathcal{N}$ as

$$c_{ij}^{\uparrow} := \sum_{m \in M_j} \frac{\gamma_{mi}^{\uparrow}}{\hat{\gamma}_{mj}} \ell_m^{\uparrow} . \tag{3.3}$$

We can thus eliminate the individual user powers and reformulate (3.1):

$$\begin{aligned} \bar{p}_i^{\uparrow} &= \sum_{m \in M_i} \gamma_{mi}^{\uparrow} \alpha_m^{\uparrow} p_m^{\uparrow} + \sum_{j \neq i} \sum_{m \in M_j} \gamma_{mi}^{\uparrow} \alpha_m^{\uparrow} p_m^{\uparrow} + \eta_i^{\uparrow} \\ &= c_{ii}^{\uparrow} \bar{p}_i^{\uparrow} + \sum_{j \neq i} c_{ij}^{\uparrow} \bar{p}_j^{\uparrow} + \eta_i^{\uparrow} . \end{aligned} \tag{3.4}$$

This equation expresses that the total power received at cell i is composed of intra-cell interference, inter-cell interference, and noise.

We define the *uplink interference-coupling matrix* $\mathbf{C}^{\uparrow} := (c_{ij}^{\uparrow})_{i,j}$ and group all equations (3.4) in the *uplink interference-coupling equation system* satisfied by the received cell powers of any solution to the uplink system model:

$$\bar{\boldsymbol{p}}^{\uparrow} = \mathbf{C}^{\uparrow} \cdot \bar{\boldsymbol{p}}^{\uparrow} + \boldsymbol{\eta}^{\uparrow} . \tag{3.5}$$

The coupling matrix is nonnegative and the noise vector is strictly positive. Furthermore, the constraints (2.7d) in the system model imply that cell powers are nonnegative, so we can determine the solvability of (2.7):

Lemma 3.2: *Consider an arbitrary radio network and an arbitrary E_c-covered snapshot and let $\mathbf{C}^{\uparrow}$ be the associated coupling matrix.*
(a) If $\rho(\mathbf{C}^{\uparrow}) \geq 1$, then no solution to the uplink system model (2.7) exists.

Figure 3.1: Graphical representation of an interference-coupling matrix

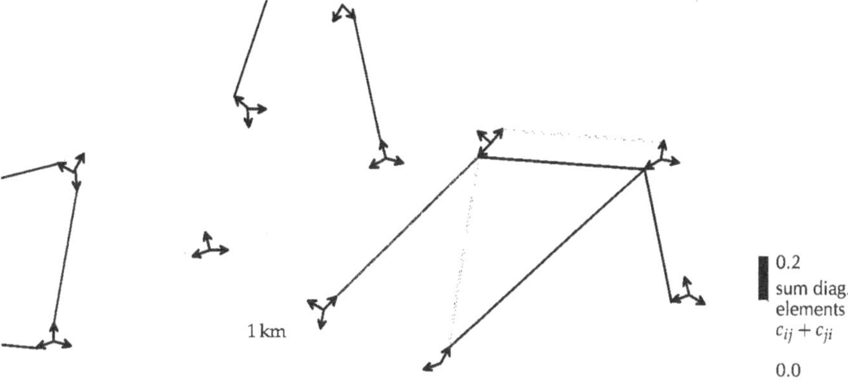

(b) If $\rho(\mathbf{C}^{\uparrow}) < 1$, Lemma 3.1 guarantees a unique solution to (3.5); we denote it by $\bar{\mathbf{p}}^{\uparrow}$. If

$$\begin{aligned} \bar{p}_i^{\uparrow} &\leq \bar{p}_{\max}^{\uparrow} && \text{for all } i \in \mathcal{N} \text{ and} \\ p_m^{\uparrow} := \bar{p}_i^{\uparrow} \ell_m^{\uparrow} / \alpha_m^{\uparrow} \gamma_{mi}^{\uparrow} &\leq p_{\max}^{(\text{UE})} && \text{for all } m \in M_i, i \in \mathcal{N}, \end{aligned} \tag{3.6}$$

then $(\bar{p}_i^{\uparrow})_{i \in \mathcal{N}}$ and $(p_m^{\uparrow})_{m \in M}$ are the unique solution to (2.7). If any of the above inequalities is violated, there is no solution to the system (2.7). □

By stating the system in this form, we aggregate all user properties and focus on the relations between cells. If uplink coverage is neglected, then we can solve the coupling equation system to determine the cell powers without computing individual link powers. In urban networks with hundreds of cells and thousand of users, this is a computational advantage. Fig. 3.1 depicts an example coupling matrix corresponding to the user configuration in Fig. 2.6.

Downlink coupling equation system. For a solution to the downlink system model (2.8), the total average transmit power of cell i as specified in (2.8b) is calculated as the sum of the average powers on dedicated links and the power on common channels

$$\bar{p}_i^{\downarrow} = \sum_{m \in M_i} \alpha_m^{\downarrow} p_{im}^{\downarrow} + p_i^{(\text{c})} . \tag{3.7}$$

For each mobile $m \in M_i$, the CIR equality (2.8a) holds:

$$\gamma_{im}^{\downarrow} p_{im}^{\downarrow} = \mu_m^{\downarrow} \left(\gamma_{im}^{\downarrow} \omega_m \left(\bar{p}_i^{\downarrow} - \alpha_m^{\downarrow} p_{im}^{\downarrow} \right) + \textstyle\sum_{j \neq i} \gamma_{jm}^{\downarrow} \bar{p}_j^{\downarrow} + \eta_m^{\downarrow} \right) .$$

Solving the CIR equation for the element $p_{im}^{\downarrow} \alpha_m^{\downarrow}$ in the sum in (3.7), we obtain

$$p_{im}^{\downarrow} \alpha_m^{\downarrow} = \frac{\alpha_m^{\downarrow} \mu_m^{\downarrow}}{1 + \omega_m \alpha_m^{\downarrow} \mu_m^{\downarrow}} \left(\omega_m \bar{p}_i^{\downarrow} + \sum_{j \neq i} \frac{\gamma_{jm}^{\downarrow}}{\gamma_{im}^{\downarrow}} \bar{p}_j^{\downarrow} + \frac{\eta_m^{\downarrow}}{\gamma_{im}^{\downarrow}} \right) . \tag{3.8}$$

We extract the user parameters in this equation into the *downlink user loading factor*

$$\ell_m^{\downarrow} := \frac{\alpha_m^{\downarrow}\,\mu_m^{\downarrow}}{1+\omega_m\alpha_m^{\downarrow}\,\mu_m^{\downarrow}} \geq 0 \tag{3.9}$$

and use this quantity when substituting (3.8) into (3.7) to obtain an alternative expression of the total transmit power:

$$\bar{p}_i^{\downarrow} = \sum_{m\in M_i}\Big(\ell_m^{\downarrow}\omega_m\,\bar{p}_i^{\downarrow} + \sum_{j\neq i}\ell_m^{\downarrow}\frac{\gamma_{jm}^{\downarrow}}{\gamma_{im}^{\downarrow}}\,\bar{p}_j^{\downarrow} + \ell_m^{\downarrow}\frac{\eta_m^{\downarrow}}{\gamma_{im}^{\downarrow}}\Big) + p_i^{(c)}\,. \tag{3.10}$$

We define the downlink *interference-coupling coefficients* $(c^{\downarrow})_{i,j\in\mathcal{N}}$ for $i,j\in\mathcal{N}$, $i\neq j$ as

$$c_{ii}^{\downarrow} := \sum_{m\in M_i}\omega_m\ell_m^{\downarrow}\,, \qquad c_{ij}^{\downarrow} := \sum_{m\in M_i}\frac{\gamma_{jm}^{\downarrow}}{\gamma_{im}^{\downarrow}}\ell_m^{\downarrow}\,, \tag{3.11}$$

and the *downlink noise power* as

$$p_i^{(\eta)} := \sum_{m\in M_i}\frac{\eta_m^{\downarrow}}{\gamma_{im}^{\downarrow}}\ell_m^{\downarrow}\,. \tag{3.12}$$

Substituting these new shorthands into (3.10) leads to a compact analog to (3.4) for the downlink:

$$\bar{p}_i^{\downarrow} = c_{ii}^{\downarrow}\,\bar{p}_i^{\downarrow} + \sum_{j\neq i}c_{ij}^{\downarrow}\bar{p}_j^{\downarrow} + p_i^{(\eta)} + p_i^{(c)}\,. \tag{3.13}$$

This equation expresses that a cell spends transmit power for overcoming intra-cell interference, inter-cell interference, and noise; furthermore, it transmits on common channels. We define the *downlink interference-coupling matrix* as $\mathbf{C}^{\downarrow} := (c_{ij}^{\downarrow})_{i,j}$ and obtain the *downlink interference-coupling system* in vector notation:

$$\bar{\boldsymbol{p}}^{\downarrow} = \mathbf{C}^{\downarrow}\bar{\boldsymbol{p}}^{\downarrow} + \boldsymbol{p}^{(\eta)} + \boldsymbol{p}^{(c)}\,. \tag{3.14}$$

In practice, for instance in urban settings, noise is often negligible because $\boldsymbol{p}^{(\eta)} \ll \boldsymbol{p}^{(c)}$.

In this context, it is also convenient to define the *downlink traffic loading factor* as

$$\tau_i^{\downarrow} := \sum_{m\in M_i}\ell_m^{\downarrow}\,. \tag{3.15}$$

This quantity represents the offered traffic from a user perspective, who is oblivious of orthogonality.

Again, the solutions to the downlink system model are necessarily nonnegative, so according to Lemma 3.1 solutions are unique and we can characterize their existence:

Lemma 3.3: *Consider an arbitrary radio network and an arbitrary E_c-covered snapshot and let $\mathbf{C}^{\downarrow}$ be the associated downlink coupling matrix.*

(a) If $\rho(\mathbf{C}^{\downarrow}) \geq 1$, then no solution to the downlink model (2.8) exists.

(b) *If $\rho(\mathbf{C}^{\downarrow}) < 1$, Lemma 3.1 guarantees a unique solution to (3.14). We denote it by $\bar{\mathbf{p}}^{\downarrow}$. If either $\bar{p}_i^{\downarrow} > \bar{p}_{\max}^{\downarrow}$ for any $i \in \mathcal{N}$, or if the E_c/I_0-coverage condition (2.8c) is violated for any $m \in M_i$, $i \in \mathcal{N}$, then the system (2.8) is infeasible. Otherwise, $(\bar{p}_i^{\downarrow})_{i \in \mathcal{N}}$ and $(p_{im}^{\downarrow})_{m \in M_i, i \in \mathcal{N}}$ defined as*

$$p_{im}^{\downarrow} := \ell_m^{\downarrow}(\omega_m \bar{p}_i^{\downarrow} + \textstyle\sum_{j \neq i} \frac{\gamma_{jm}^{\downarrow}}{\gamma_{im}^{\downarrow}} \bar{p}_j^{\downarrow} + \frac{\eta_m^{\downarrow}}{\gamma_{im}^{\downarrow}}) / \alpha_m^{\downarrow}, \qquad m \in M_i,\ i \in \mathcal{N}, \tag{3.16}$$

are the unique solution to (2.8). □

Properties of the interference-coupling matrices. Unfortunately, the coupling matrices have little structure; in particular, they are in general asymmetric. Their only reliable property is non-negativity. Mobiles are usually assigned to the server with the least attenuation (the "best server"). In this case, $\gamma_{mi}^{\uparrow}/\gamma_{mj}^{\uparrow} \leq 1$, so the diagonal element of the uplink matrix is largest in each column. However, this assignment of mobiles to cells is not mandatory, and in downlink the effect of orthogonality may prevent a similar row-wise dominance of the diagonal. In a reasonable network design, the largest part of the planning area consists of cell centers rather than border regions, so matrices are typically diagonally dominant in practice.

Because a typical cell has only a limited number of influential neighbors, there are many small entries in the coupling matrices occurring in practice. These are usually rounded to zero for practical purposes. The resulting sparsity pattern, however, has no special structure. In particular, the matrices of networks with a contiguous coverage area are usually primitive (they have a power with only positive elements); this basically means that any cell is at least indirectly coupled to any other cell.

3.2 Perfect load control and complementarity systems

The interference-coupling equations serve for efficiently calculating the solution to the complete system model—if it exists. If a snapshot cannot be served, the solutions of the coupling system do not relate to a solution of the system model. We will now generalize the coupling equations to a new system model whose solutions always indicate a solution to the classical system model. In particular, we aim at finding a solution that does not violate the cell power limits.

3.2.1 Deduction of complementarity system for an isolated uplink cell

We will motivate and illustrate our approach for the simplified setting of an isolated uplink cell first. The principle is then generalized to coupled systems and the downlink in the next section.

Pole equation and and soft capacity. For an isolated cell i, the coupling equation (3.4) has a positive solution if and only if the coupling element is smaller than one. In this

Figure 3.2: The classical uplink pole equation

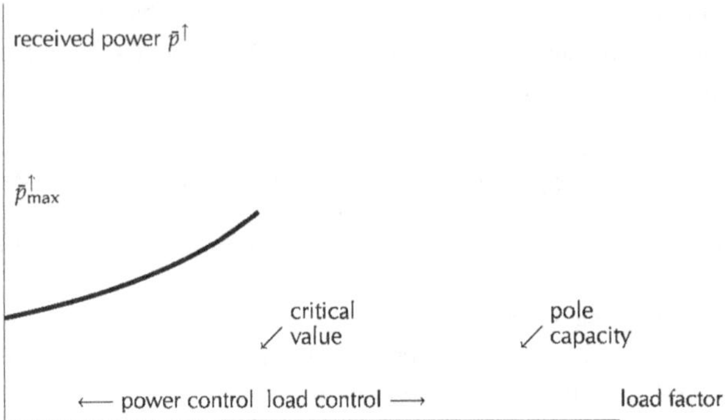

case, the cell power can be calculated as a function of $c^{\uparrow}_{ii}$ as

$$\bar{p}^{\uparrow}_i = \frac{\eta^{\uparrow}_i}{1 - c^{\uparrow}_{ii}}. \tag{3.17}$$

This function is depicted in Fig. 3.2. The traffic load corresponding to $c_{ii} = 1$ is called the system's *pole capacity;* even with infinite cell power, no more traffic can be served. Eqn. (3.17) is a *pole equation* (for an isolated cell). Inter-cell interference reduces a cell's pole capacity; a generalized pole equation for coupled cells is introduced in Sec. 3.3.

In a real system, cell power limits apply, which are ensured by load control. The pole equation hence only has a physical interpretation if the resulting power does not exceed the limit $\bar{p}^{\uparrow}_{\text{max}}$. We call the traffic load $1 - \eta^{\uparrow}_i / \bar{p}^{\uparrow}_{\text{max}}$ corresponding to the maximum feasible cell power the *critical value.* If the offered traffic load exceeds the critical value, load control kicks in and only a fraction of the users is granted access to the network. We denote the load factor after load control by $\tilde{c}^{\uparrow}_{ii} \leq 1 - \eta^{\uparrow}_i / \bar{p}^{\uparrow}_{\text{max}}$. For the reduced set of users, the pole equation describes the cell power level resulting from power control.

Perfect fractional load control and complementarity. We use the ratio of the reduced load factor $\tilde{c}^{\uparrow}_{ii}$ over the original load factor $c^{\uparrow}_{ii}$ as a measure for the impact of load control and denote it by

$$\lambda^{\uparrow}_i := \tilde{c}^{\uparrow}_{ii} / c^{\uparrow}_{ii}.$$

The cell load is then described by a revised pole equation:

$$\bar{p}^{\uparrow}_i = \lambda^{\uparrow}_i c^{\uparrow}_{ii} \bar{p}^{\uparrow}_i + \eta^{\uparrow}_i. \tag{3.18}$$

It is desirable for the network operator that as few users as possible are turned down, so $\lambda^{\uparrow}_i$ should be maximal. To ensure that the maximum cell power is not surpassed,

Figure 3.3: Complementarity in the generalized pole equation for an isolated uplink cell

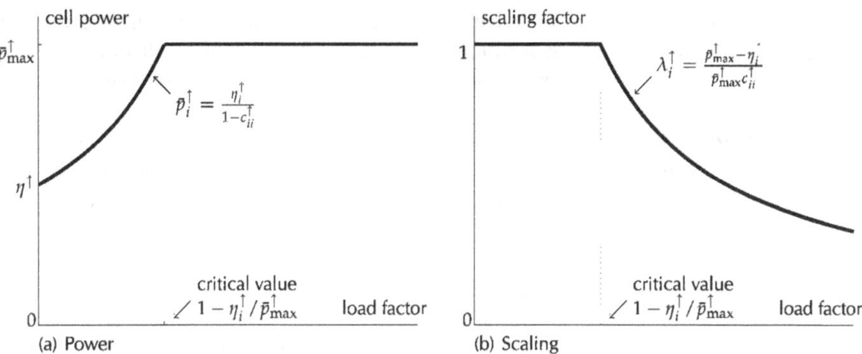

however, it is required that

$$\lambda^{\uparrow}_i \leq \frac{\bar{p}^{\uparrow}_{\max} - \eta^{\uparrow}_i}{\bar{p}^{\uparrow}_{\max} c^{\uparrow}_{ii}}. \tag{3.19}$$

The fundamental assumption of *perfect load control* is that the offered traffic is minimally reduced. This means that if the right-hand side in (3.19) is smaller than one, it should be met with equality. In this case, the cell power is precisely $\bar{p}^{\uparrow}_{\max}$. In other words, we demand *complementarity* of the two constraints

$$\bar{p}^{\uparrow}_i \leq \bar{p}^{\uparrow}_{\max} \quad \text{and} \quad \lambda^{\uparrow}_i \leq 1. \tag{3.20}$$

Complementarity here means that only one of the two inequalities may be strict.

The assumption is idealizing because it is in general not possible to precisely meet the power constraint by blocking some users. Perfect load control has a physical interpretation, if we assume that users can receive fractional service. Then either the last admitted user is served only partly, or all users are served only with a fraction $\lambda^{\uparrow}_i$ of their original traffic demand $\ell^{\uparrow}_m$. The last interpretation carries over to the coupled multi-cell case.

In the simplified single-cell case, we can hence describe the network's behavior for any snapshot by the combination of the revised pole equation (3.18) and complementarity in (3.20); the concept is illustrated in Fig. 3.3. If $c^{\uparrow}_i \leq 1 - \eta^{\uparrow}_i / \bar{p}^{\uparrow}_{\max}$, all users can be served ($\lambda^{\uparrow}_i = 1$) and the cell power is calculated according to (3.17) obeying perfect power control. Otherwise, the cell is maximally loaded ($\bar{p}^{\uparrow}_i = \bar{p}^{\uparrow}_{\max}$) and the fraction $\lambda^{\uparrow}_i$ of traffic that is admitted is calculated assuming equality in (3.19) according to perfect load control.

3.2.2 Perfect load control in a coupled system

We generalize the idea introduced in the simplified setting by formulating revised coupling equations containing additional scaling variables for each cell and imposing complementarity constraints on the scaling and power variables of each cell.

Revised coupling equations. The load control decision for cell j in uplink is represented by the scaling factor $\lambda_j^\uparrow \in (0,1]$. The factor applies to all mobiles in cell j, *i. e.*, to all uplink loading factors $\ell_m^\uparrow$ with $m \in M_j$. The *columns* of the coupling matrix are thus scaled, and the revised pendant of the coupling equations (3.4) for cell i is:

$$\bar{p}_i^\uparrow = \textstyle\sum_j \lambda_j^\uparrow c_{ij}^\uparrow \bar{p}_j^\uparrow + \eta_i^\uparrow \,. \tag{3.21}$$

The load control decision for cell i in downlink is represented by the scaling factor $\lambda_i^\downarrow \in [0,1]$. The sum (3.10) only contains loading factors $\ell_m^\downarrow$ of mobiles in cell i. Accordingly, the scaling factor $\lambda_i^\downarrow$ applies to the *row* of the coupling matrix. In addition, the i-th component of the noise load vector (3.12) is affected by $\lambda_i^\downarrow$. The revised version of (3.13) hence reads

$$\bar{p}_i^\downarrow = \lambda_i^\downarrow (c_{ii}^\downarrow \bar{p}_i^\downarrow + \textstyle\sum_{j \neq i} c_{ij}^\downarrow \bar{p}_j^\downarrow + p_i^{(\eta)}) + p_i^{(c)} \,. \tag{3.22}$$

In the following, we will refer the values $\lambda_i^\uparrow$ and $\lambda_i^\downarrow$ as the *scaling factor* or the *grade of service* of cell i in uplink and downlink, respectively. Occasionally, we will also use the term *blocking ratio* for the value $1 - \lambda_i$. We say that a cell i is *load-controlled* or *blocking* in uplink or downlink if $\lambda_i^\uparrow < 1$ or $\lambda_i^\downarrow < 1$.

Complementarity conditions. The central complementarity condition is similar to the single-cell case: We require that a cell is only load-controlled if it would otherwise violate the maximum admissible power value. If the cell is load-controlled, then we require that the load control decision must be taken such that the radio resource is used to the fullest. If the constraints on cell powers are imposed, this condition is for the uplink equivalent to

$$\lambda_i^\uparrow < 1 \Rightarrow \bar{p}_i^\uparrow = \bar{p}_{\max}^\uparrow \,, \tag{3.23}$$

and, analogously, we require for the downlink that

$$\lambda_i^\downarrow < 1 \Rightarrow \bar{p}_i^\downarrow = \bar{p}_{\max}^\downarrow \,. \tag{3.24}$$

Note that the complementarity conditions relate only the power and scaling variables of the same cell. A cell thus behaves greedily: it only curbs the traffic demand of its users to regulate its own load, but not to alleviate the situation of a neighbor.

Complete interference-coupling complementarity systems. Recall that we use the notation $\bar{\boldsymbol{p}}^{\uparrow}_{\max}$ for a vector containing the scalar $\bar{p}^{\uparrow}_{\max}$ in each component. The complementarity system for the uplink in vector notation is:

$$\begin{aligned} \bar{\boldsymbol{p}}^{\uparrow} &= \boldsymbol{C}^{\uparrow}\mathrm{diag}(\boldsymbol{\lambda}^{\uparrow})\,\bar{\boldsymbol{p}}^{\uparrow} + \boldsymbol{\eta}^{\uparrow} && (3.25a)\\ 0 &= (\boldsymbol{1} - \boldsymbol{\lambda}^{\uparrow})^{t}(\bar{\boldsymbol{p}}^{\uparrow}_{\max} - \bar{\boldsymbol{p}}^{\uparrow}) && (3.25b)\\ \bar{\boldsymbol{p}}^{\uparrow} &\in [\boldsymbol{\eta}^{\uparrow}, \bar{\boldsymbol{p}}^{\uparrow}_{\max}]\\ \boldsymbol{\lambda}^{\uparrow} &\in [\boldsymbol{0}, \boldsymbol{1}] \end{aligned}$$

The equation system (3.25a) is the revised coupling equation in vector form; the row (3.25b) is written out as $\sum_i (1 - \lambda_i^{\uparrow})(\bar{p}^{\uparrow}_{\max} - \bar{p}_i^{\uparrow}) = 0$. Due to the bounds on the variables, every element of the sum is nonnegative, so (3.25b) is equivalent to (3.23). For the downlink, we have:

$$\begin{aligned} \bar{\boldsymbol{p}}^{\downarrow} &= \mathrm{diag}(\boldsymbol{\lambda}^{\downarrow})\,(\boldsymbol{C}^{\downarrow}\bar{\boldsymbol{p}}^{\downarrow} + \boldsymbol{p}^{(\eta)}) + \boldsymbol{p}^{(c)} && (3.26a)\\ 0 &= (\boldsymbol{1} - \boldsymbol{\lambda}^{\downarrow})^{t}(\bar{\boldsymbol{p}}^{\downarrow}_{\max} - \bar{\boldsymbol{p}}^{\downarrow}) && (3.26b)\\ \bar{\boldsymbol{p}}^{\downarrow} &\in [\boldsymbol{p}^{(c)}, \bar{\boldsymbol{p}}^{\downarrow}_{\max}]\\ \boldsymbol{\lambda}^{\downarrow} &\in [\boldsymbol{0}, \boldsymbol{1}] \end{aligned}$$

As long as outage due to a lack of uplink coverage (2.7c) or E_c/I_0 coverage (2.8c) is disregarded, the complementarity systems generalize the classical system model when it is infeasible:

Lemma 3.4: *For any network and any E_c-covered snapshot, the following statements hold true:*

(a) *If the classical uplink system model (2.7) has a solution with cell powers $\bar{\boldsymbol{p}}^{\uparrow}$, then $(\bar{\boldsymbol{p}}^{\uparrow}, \boldsymbol{1})$ is the unique solution to the uplink complementarity system (3.25). If, on the other hand, (3.25) has a solution $(\bar{\boldsymbol{p}}^{\uparrow}, \boldsymbol{\lambda}^{\uparrow})$ with $\boldsymbol{\lambda}^{\uparrow} \neq \boldsymbol{1}$, then (2.7) is infeasible. For the snapshot with modified parameters $\tilde{\mu}^{\uparrow}_m := \mu^{\uparrow}_m \lambda^{\uparrow}_i / (1 + (1 - \lambda^{\uparrow}_i)\alpha^{\uparrow}_m \mu^{\uparrow}_m)$, $m \in M_i$, however, the cell powers $\bar{\boldsymbol{p}}^{\uparrow}$ and link powers $(p^{\uparrow}_m)_{m \in M}$ defined in (3.6) are the feasible solution to (2.7) provided that the uplink coverage conditions (2.7c) are fulfilled.*

(b) *Analogously, for a solution to (2.8) with cell powers $\bar{\boldsymbol{p}}^{\downarrow}$, $(\bar{\boldsymbol{p}}^{\downarrow}, \boldsymbol{1})$ is the unique solution to (3.26). On the other hand, any solution $(\bar{\boldsymbol{p}}^{\downarrow}, \boldsymbol{\lambda}^{\downarrow})$ to (3.26) with $\boldsymbol{\lambda}^{\downarrow} \neq \boldsymbol{1}$ rules out the feasibility of (2.8). For the modified snapshot with $\tilde{\mu}^{\uparrow}_m := \mu^{\downarrow}_m \lambda^{\downarrow}_i / (1 + (1 - \lambda^{\downarrow}_i)\alpha^{\downarrow}_m \mu^{\downarrow}_m)$ for $m \in M_i$, however, the cell powers $\bar{\boldsymbol{p}}^{\downarrow}$ and link powers $(p^{\downarrow}_{im})_{m \in M_i, i \in \mathcal{N}}$ defined in (3.16) are the unique solution to (2.8) provided that the E_c/I_0 coverage conditions (2.8c) hold.*

Sketch of proof. If $\boldsymbol{\lambda} = \boldsymbol{1}$, then the system is equivalent to the original coupling system, and Lemmas 3.2 and 3.3 apply. If there is a solution with any blocking cell, then consider $\bar{\boldsymbol{p}}$ as a function of $\boldsymbol{\lambda}$ defined by the scaled coupling equation (3.25a) and (3.26a). This function depends monotonously, and in the load-controlled components even strongly monotonously, on $\boldsymbol{\lambda}$. There is hence no way of increasing $\boldsymbol{\lambda}$ to $\boldsymbol{1}$ without violating the soft blocking constraint of the currently load-controlled cell.

The modified CIR targets $\tilde{\mu}$ are calculated such that the resulting user load factor for each user equals the scaled original user load factor. □

We thus have a compact, closed-form representation of the system including soft capacity and load control, but user-specific load-control decisions and some cases of outage are not covered. We discuss computational experiments below, which suggest that these inaccuracies are tolerable in typical realistic settings. We develop methods for deriving user-specific load-control decisions from the scaling factors in Sec. 3.2.4, and computational tests in Sec. 4.4.1 show that perfect load control is a reasonable approximation. In the optimization case studies presented in Sec. 5.4, E_c/I_0 coverage increases through optimization even though it is disregarded in the system model.

3.2.3 Solving the complementarity systems

For using the complementarity systems in practice, we need to be able to efficiently solve them. The problem resembles *linear complementarity systems* (*Berman & Plemmons*, 1994, Ch. 10), but the revised coupling relations are nonlinear in the power and scaling variables. *Chung* (1989) showed that solving linear complementarity problems is NP-hard[1] in general. It is hence not a priori clear whether solutions for our instances of nonlinear systems can be found in reasonable time.

Luckily, the structure of the downlink system is benign: we will see that there always exists a unique solution, which can be computed by an iterative scheme. For the uplink, we transform problem (3.25) into a generalized linear complementarity system and develop a heuristic.

Downlink recursion and uniqueness of solution. We assume that $\sum_j c^{\downarrow}_{ij} + p^{(\eta)}_i > 0$ for all $i \in \mathcal{N}$. (If any cell is empty, the scaling value is one and the cell can be excluded.) The following iterative scheme allows to find a solution to (3.26):

$$\tilde{\lambda}^{(0)}_i := 1, \qquad \bar{p}^{(0)}_i := p^{(c)}_i,$$

$$\tilde{\lambda}^{(t+1)}_i := \min\left(\tilde{\lambda}^{(t)}_i, \frac{\bar{p}^{\downarrow}_{\max} - p^{(c)}_i}{c^{\downarrow}_{ii}\bar{p}^{\downarrow}_{\max} + p^{(\eta)}_i + \sum_{j\neq i} c^{\downarrow}_{ij}\bar{p}^{(t)}_j}\right), \tag{3.27a}$$

$$\bar{p}^{(t+1)}_i := \frac{p^{(c)}_i + \tilde{\lambda}^{(t+1)}_i\left(p^{(\eta)}_i + \sum_{j\neq i} c^{\downarrow}_{ij}\bar{p}^{(t)}_j\right)}{1 - \tilde{\lambda}^{(t+1)}_i c^{\downarrow}_{ii}}. \tag{3.27b}$$

The iteration is identical to the Jacobi method (*Berman & Plemmons*, 1994) for solving the linear coupling equation (3.14), if the update (3.27a) for the scaling variable is omitted. Taking the minimum in (3.27a) ensures that the power update does not result in a negative power value or in a value exceeding $\bar{p}^{\downarrow}_{\max}$. In this case, there is a new minimum in (3.27a), and the scaling value is adjusted such that the next power update (3.27b) results in $\bar{p}^{(t)}_i = \bar{p}^{\downarrow}_{\max}$. Otherwise a standard Jacobi step is performed.

[1] An introduction to complexity theory and a formal definition of NP hardness can be found, for example, in the book by *Cormen et al.* (1990); briefly, the term means that an efficient algorithm for solving arbitrary instances is unlikely to exist.

Lemma 3.5: *The sequences* $(\tilde{\lambda}_i^{(t)})_{t\geq 0}$ *and* $(\tilde{p}_i^{(t)})_{t\geq 0}$ *converge to the unique solution of the downlink complementarity system* (3.26).

Proof. We first check convergence. Induction shows that the updates (3.27b) for $\tilde{p}_i^{(t)}$ are monotonously increasing for growing t. The scaling values are monotonously decreasing because of the minimum in (3.27a). Furthermore, both sequences are bounded, so we have:

$$1 \geq \tilde{\lambda}_i^{(t)} \geq \tilde{\lambda}_i^{(t+1)} \geq 0 \qquad \text{for all } t \geq 0,\ i \in \mathcal{N},$$
$$p_i^{(c)} = \tilde{p}_i^{(t)} \leq \tilde{p}_i^{(t+1)} \leq \bar{p}_{\max}^{\downarrow} \qquad \text{for all } t \geq 0,\ i \in \mathcal{N}.$$

Hence, both sequences converge.

The limits $\lim_{t\to\infty} \tilde{\boldsymbol{\lambda}}^{(t)}$ and $\lim_{t\to\infty} \tilde{\boldsymbol{p}}^{(t)}$ fulfill (3.26a); this is obvious from (3.27b). To be a solution to the complete system (3.26), the limits have to obey the complementarity complementarity (3.26b), which is equivalent to (3.24). The implication

$$\tilde{\lambda}_i^{(t)} < 1 \Rightarrow \tilde{p}_i^{(t)} = \bar{p}_{\max}^{\downarrow}$$

holds by construction for all i and all t. This can be seen by substituting the right-hand term in the minimum in (3.27a) into (3.27b). With the above monotonicity, complementarity is thus also ensured for the limits, and they are a solution to the system (3.26).

We last show uniqueness. Suppose that both $(\bar{\boldsymbol{p}}^{(1)}, \boldsymbol{\lambda}^{(1)})$ and $(\bar{\boldsymbol{p}}^{(2)}, \boldsymbol{\lambda}^{(2)})$ are solutions of (3.26). Then for $\bar{\boldsymbol{p}}^{(\min)} := \min(\bar{\boldsymbol{p}}^{(1)}, \bar{\boldsymbol{p}}^{(2)})$ and $\boldsymbol{\lambda}^{(\max)} := \max(\boldsymbol{\lambda}^{(1)}, \boldsymbol{\lambda}^{(2)})$, where the maximum and minimum are understood componentwise, we can establish that

$$\bar{\boldsymbol{p}}^{(\min)} \geq \operatorname{diag}(\boldsymbol{\lambda}^{(\max)}) \left(\boldsymbol{C}^{\downarrow} \bar{\boldsymbol{p}}^{(\min)} + \boldsymbol{p}^{(\eta)}\right) + \boldsymbol{p}^{(c)}. \tag{3.28}$$

To see this, consider an index $i \in \mathcal{N}$. We distinguish two cases:

(a) If $\bar{p}_i^{(1)} \neq \bar{p}_i^{(2)}$, we may assume that $\bar{p}_i^{(1)} < \bar{p}_i^{(2)}$. Then by complementarity $\lambda_i^{(1)} = \lambda_i^{(\max)} = 1$. Then

$$\bar{p}_i^{(2)} > \bar{p}_i^{(1)} = \textstyle\sum_j c_{ij}^{\downarrow} \bar{p}_j^{(1)} + p_i^{(\eta)} + p_i^{(c)} \geq \lambda_i^{(\max)} \left(\sum_j c_{ij}^{\downarrow} \bar{p}_j^{(\min)} + p_i^{(\eta)}\right) + p_i^{(c)}.$$

(b) If $\bar{p}_i^{(1)} = \bar{p}_i^{(2)}$, we assume that $\lambda_i^{(1)} \leq \lambda_i^{(2)}$, hence $\lambda_i^{(2)} = \lambda_i^{(\max)}$. Then

$$\bar{p}_i^{(1)} = \bar{p}_i^{(2)} = \lambda_i^{(2)} \left(\textstyle\sum_j c_{ij}^{\downarrow} \bar{p}_j^{(2)} + p_i^{(\eta)}\right) + p_i^{(c)} \geq \lambda_i^{(\max)} \left(\sum_j c_{ij}^{\downarrow} \bar{p}_j^{(\min)} + p_i^{(\eta)}\right) + p_i^{(c)}.$$

Using, for example, the arguments from *Imhof & Mathar* (2005, Lemma 1), we can deduce from (3.28) that there is a (unique) vector $\bar{\boldsymbol{p}}^{(0)}$ with $\boldsymbol{p}^{(c)} \leq \bar{\boldsymbol{p}}^{(0)} \leq \bar{\boldsymbol{p}}^{(\min)}$ with

$$\bar{\boldsymbol{p}}^{(0)} = \operatorname{diag}(\boldsymbol{\lambda}^{(\max)}) \left(\boldsymbol{C}^{\downarrow} \bar{\boldsymbol{p}}^{(0)} + \boldsymbol{p}^{(\eta)}\right) + \boldsymbol{p}^{(c)}.$$

This shows that $\rho(\operatorname{diag}(\boldsymbol{\lambda}^{(\max)})\, \boldsymbol{C}^{\downarrow}) < 1$. The solution of (3.26a) (for $\bar{\boldsymbol{p}}^{\downarrow}$) hence exists and is positive for all $\boldsymbol{\lambda}^{\downarrow} \in [\boldsymbol{0}, \boldsymbol{\lambda}^{(\max)}]$. Moreover, this solution depends strongly monotonously on $\boldsymbol{\lambda}^{\downarrow}$. As $\boldsymbol{\lambda}^{(\max)} \geq \boldsymbol{\lambda}^{(1)}$, we would therefore also have $\bar{\boldsymbol{p}}^{(0)} \geq \bar{\boldsymbol{p}}^{(1)}$, and analogously $\bar{\boldsymbol{p}}^{(0)} \geq \bar{\boldsymbol{p}}^{(2)}$. Then the only remaining option is $\bar{\boldsymbol{p}}^{(0)} = \bar{\boldsymbol{p}}^{(1)} = \bar{\boldsymbol{p}}^{(2)}$ and $\boldsymbol{\lambda}^{(\max)} = \boldsymbol{\lambda}^{(1)} = \boldsymbol{\lambda}^{(2)}$. □

The Gauß-Seidel method (*Berman & Plemmons*, 1994) is a variant of the Jacobi method for solving linear equation systems, which often converges faster; it differs in that updated solution approximates are used already in the current iteration. The convergence of the Jacobi and standard Gauß-Seidel methods depends on the structure of the matrix. Convergence is basically the faster the closer the matrix is to a diagonal matrix; this property seems to carry over to our iterative scheme as well. In practice, the method converges quickly; a theoretical analysis is open yet.

Number of solutions in uplink. In the uplink, the number of solutions to (3.25) depends on the coupling matrix; there might be no solution, one solution, and even infinitely many. Consider, the following simple examples with parameters $\bar{p}^{\uparrow}_{\max} = 2$, $\eta^{\uparrow} = \binom{1}{1}$. For $C^{\uparrow} = \left(\begin{smallmatrix}1 & 0\\ 2 & 1\end{smallmatrix}\right)$ there is no solution; complementarity in the first cell requires $\lambda^{\uparrow}_1 = 1/2$, and with this value we have $\bar{p}^{\uparrow}_2 \geq 3$, *i.e.*, the second cell violates soft capacity constraints. For $C^{\uparrow} = \left(\begin{smallmatrix}1 & 0\\ 0 & 1\end{smallmatrix}\right)$ the system decomposes, and the solution $\boldsymbol{\lambda}^{\uparrow} = \binom{1/2}{1/2}$, $\bar{\boldsymbol{p}}^{\uparrow} = \binom{2}{2}$ is unique. For $C^{\uparrow} = \left(\begin{smallmatrix}1 & 1\\ 1 & 1\end{smallmatrix}\right)$, there are infinitely many solutions; picking $\boldsymbol{\lambda}^{\uparrow}$ as any convex combination of $\binom{1/2}{0}$ and $\binom{0}{1/2}$ produces a solution together with $\bar{\boldsymbol{p}}^{\uparrow} = \binom{2}{2}$. There are also examples of (finitely many) multiple solutions for an asymmetrical coupling matrix exhibiting the typical column-wise diagonal dominance (*Eisenblätter et al.*, 2005*b*).

The above example of an infeasible system has an intuition: In an unfavorable network design, the neighbors may jam an empty cell with interference. To model this event, the classical uplink system (2.7) would have to be altered to allow a violation of the uplink cell power in empty cells. In consequence, the power limits would be abolished in (3.25) if the scaling value is zero. Because this case is foremost of theoretical interest and leads to technical case decisions in the following, we have omitted it and opted for better clarity of presentation. The variant of complementarity systems including empty cells is described in an earlier paper (*Eisenblätter et al.*, 2005*b*). A characterization of the number of solutions of the uplink system system is open.

Linearization of the uplink system. The problem (3.25) can be linearized and transformed into an extended linear complementarity problem (*De Schutter & De Moor*, 1995). To this end, we introduce an auxiliary variable

$$\dot{p}^{\uparrow}_i = \lambda^{\uparrow}_i \bar{p}^{\uparrow}_i$$

for each cell i. Since $\lambda^{\uparrow}_i \leq 1$, we have $\dot{p}^{\uparrow}_i \leq \bar{p}^{\uparrow}_i$. By enforcing complementarity between this inequality and the upper bound for $\bar{p}^{\uparrow}_i$, we obtain the following problem:

$$\begin{aligned}
&\bar{\boldsymbol{p}}^{\uparrow} = C^{\uparrow}\dot{\boldsymbol{p}}^{\uparrow} + \eta^{\uparrow} \\
&\dot{\boldsymbol{p}}^{\uparrow} \leq \bar{\boldsymbol{p}}^{\uparrow} \qquad\qquad \perp \quad \bar{\boldsymbol{p}}^{\uparrow} \leq \bar{\boldsymbol{p}}^{\uparrow}_{\max} \\
&\mathbf{0} \leq \dot{\boldsymbol{p}}^{\uparrow}
\end{aligned} \tag{3.29}$$

The relation $\perp$ here applies, independently, to all components of the vectors and encodes the requirement that at most one of the two inequalities may have a slack. The direct correspondence between feasible solutions to (3.29) and (3.25) is obvious. As the slacks in all inequalities are bounded, we can transform problem (3.29) to a mixed integer feasibility problem using the methods from *De Schutter et al.* (2002). To this end, two additional binary variables per pair of inequalities enforce complementarity:

Lemma 3.6: *For any coupling matrix $\mathbf{C}^\uparrow \in \mathbb{R}_+^{\mathcal{N}\times\mathcal{N}}$, the solutions of the complementarity system* (3.25) *correspond to the solutions of the mixed integer linear feasibility problem*

$$\begin{aligned}
&\bar{\boldsymbol{p}}^\uparrow = \mathbf{C}^\uparrow \dot{\boldsymbol{p}}^\uparrow + \boldsymbol{\eta}^\uparrow \\
&\mathbf{0} \leq \bar{\boldsymbol{p}}^\uparrow - \dot{\boldsymbol{p}}^\uparrow \leq \operatorname{diag}(\bar{\boldsymbol{p}}^\uparrow_{\max})\,\boldsymbol{\theta} \\
&\mathbf{0} \leq \bar{\boldsymbol{p}}^\uparrow_{\max} - \bar{\boldsymbol{p}}^\uparrow \leq \operatorname{diag}(\bar{\boldsymbol{p}}^\uparrow_{\max})\,\boldsymbol{\zeta} \\
&\boldsymbol{\theta} + \boldsymbol{\zeta} \leq \mathbf{1} \\
&\dot{\boldsymbol{p}}^\uparrow, \bar{\boldsymbol{p}}^\uparrow \geq \mathbf{0} \\
&\boldsymbol{\theta}, \boldsymbol{\zeta} \in \{0,1\}^n
\end{aligned} \tag{3.30}$$

in the following sense:

(a) For any feasible solution $(\bar{\boldsymbol{p}}^\uparrow, \boldsymbol{\lambda}^\uparrow)$ of the system (3.25)*, set $\dot{p}_i^\uparrow := \lambda_i^\uparrow \bar{p}_i^\uparrow$. Then there are binary vectors $\boldsymbol{\theta}, \boldsymbol{\zeta} \in \{0,1\}^n$ such that* (3.30) *is feasible.*

(b) For any feasible solution of (3.30)*, set $\lambda_i^\uparrow := \dot{p}_i^\uparrow / \bar{p}_i^\uparrow$. Then the pair $(\bar{\boldsymbol{p}}^\uparrow, \boldsymbol{\lambda}^\uparrow)$ is a solution to* (3.25). □

An extended mixed integer linear feasibility model including the case of empty overloaded cells is described in (*Eisenblätter et al.*, 2005*b*).

Heuristic uplink recursion. We specify a heuristic algorithm for solving (3.25); it is an iterative scheme inspired by the downlink iteration (3.27). The additional difficulty is that an estimate for cell power and scaling of a given cell cannot be computed by considering (estimates for) the powers of other cells only, but estimates for their traffic scaling factors need to be factored in as well. We hence use two estimates $\hat{\lambda}_i^{(t)}, \check{\lambda}_i^{(t)}$ for the scaling and $\hat{p}_i^{(t)}, \check{p}_i^{(t)}$ for the power values that represent upper and lower bounds:

$$\check{p}_i^{(0)} := \eta_i^\uparrow, \qquad \hat{p}_i^{(0)} := \bar{p}^\uparrow_{\max}, \qquad \check{\lambda}_i^{(0)} := 0, \qquad \hat{\lambda}_i^{(0)} := 1, \tag{3.31a}$$

$$\hat{\lambda}_i^{(t+1)} := \max\Big(0, \min\Big(\hat{\lambda}_i^{(t)}, (\bar{p}^\uparrow_{\max} - \textstyle\sum_{j\neq i} \check{\lambda}_j^{(t)} c_{ij}^\uparrow \check{p}_j^{(t)} - \eta_i^\uparrow) / \bar{p}^\uparrow_{\max} c_{ii}^\uparrow\Big)\Big), \tag{3.31b}$$

$$\check{\lambda}_i^{(t+1)} := \min\Big(1, \max\Big(\check{\lambda}_i^{(t)}, (\bar{p}^\uparrow_{\max} - \textstyle\sum_{j\neq i} \hat{\lambda}_j^{(t)} c_{ij}^\uparrow \hat{p}_j^{(t)} - \eta_i^\uparrow) / \bar{p}^\uparrow_{\max} c_{ii}^\uparrow\Big)\Big), \tag{3.31c}$$

$$\hat{p}_i^{(t+1)} := \frac{\sum_{j\neq i} \hat{\lambda}_j^{(t+1)} c_{ij}^\uparrow \hat{p}_j^{(t)} + \eta_i^\uparrow}{1 - \hat{\lambda}_i^{(t+1)} c_{ii}^\uparrow}, \tag{3.31d}$$

$$\check{p}_i^{(t+1)} := \frac{\sum_{j\neq i} \check{\lambda}_j^{(t+1)} c_{ij}^\uparrow \check{p}_j^{(t)} + \eta_i^\uparrow}{1 - \check{\lambda}_i^{(t+1)} c_{ii}^\uparrow}. \tag{3.31e}$$

If a cell is empty and $c_{ii}^\uparrow = 0$, it does not influence the cell powers in other cells and is excluded from the iteration; its power value may be computed afterwards. Note that the denominators in (3.31d) and (3.31e) are always adjusted such that the result is well defined and positive.

Lemma 3.7: *The series* $(\hat{\lambda}_i^{(t)})_{t\geq 0}$, $(\check{\lambda}_i^{(t)})_{t\geq 0}$, $(\hat{p}_i^{(t)})_{t\geq 0}$, *and* $(\check{p}_i^{(t)})_{t\geq 0}$ *converge. If*

$$\lim_{t\to\infty} \hat{\lambda}_i^{(t)} = \lim_{t\to\infty} \check{\lambda}_i^{(t)} \qquad \text{and} \qquad \lim_{t\to\infty} \hat{p}_i^{(t)} = \lim_{t\to\infty} \check{p}_i^{(t)} \leq \bar{p}_{\max}^{\uparrow} \qquad \text{for all } i \in \mathcal{N},$$

then the limits are a solution to the uplink complementarity system (3.25).

Proof. The sequences $\hat{\lambda}_i^{(t)}$ and $\check{\lambda}_i^{(t)}$ are monotonous and bounded by construction, and therefore converge. In consequence, also the sequences $\hat{p}_i^{(t)}$ and $\check{p}_i^{(t)}$ converge. Moreover, the pairs $(\hat{\lambda}_i^{(t)}, \check{p}_i^{(t)})$ and $(\check{\lambda}_i^{(t)}, \hat{p}_i^{(t)})$ are (by construction) complementary in the sense of (3.23) if the soft capacity constraint holds. If the series converge to a common limit, then the limits are a solution of (3.25a) by construction. □

The same proof works for various modifications of the above scheme, but this version was observed to have best convergence properties for realistic instances. In all experiments based on realistic data conducted for this thesis, the iteration (3.31) converged.

There is, however, no guarantee for convergence in general; the scheme thus remains a heuristic. In the above example of an infeasible system with $\mathbf{C}^{\uparrow} = \begin{pmatrix} 1 & 0 \\ 2 & 1 \end{pmatrix}$, the iterates stall at $\hat{\boldsymbol{\lambda}}^{(2)} = \check{\boldsymbol{\lambda}}^{(2)} = \binom{1/2}{0}$, $\hat{\boldsymbol{p}}^{(2)} = \binom{2}{2}$, and $\check{\boldsymbol{p}}^{(2)} = \binom{2}{3}$. On the other hand, if the coupling matrix is a diagonal matrix, it is easy to see that convergence to the unique solution is obtained in the first iteration. As discussed above, realistic instances of coupling matrices are typically diagonally dominant; this might be the reason for the practical success of the heuristic. A characterization of the instances for which convergence takes place remains open.

3.2.4 Discrete load control decisions

With the complementarity systems, we can abstract from individual users and understand a radio network at cell level. We use only this top-level view in the following. To validate our underlying assumption of perfect load control, however, we need find out how close it matches the traditional model of user-based load control. To this end, we describe methods for translating the scaling values of a solution of a complementarity system back to a binary load-control decision per user. As a secondary application, our proposals may be used to find a good starting solution for traditional user-based load-control schemes.

We wish to determine a subset of served users $M' \subseteq M$, thereby rounding the fractional scaling factors a value in $\{0,1\}$ for each user (1 corresponding to including the user in M'). We write $\mathbf{C}^{\uparrow}, \mathbf{C}^{\downarrow}$ and $\tilde{\mathbf{C}}^{\uparrow} = (\tilde{c}_{ij}^{\uparrow})_{i,j}$, $\tilde{\mathbf{C}}^{\downarrow} = (\tilde{c}_{ij}^{\downarrow})_{i,j}$ for the uplink and downlink coupling matrices before and after removing users; $\tilde{\boldsymbol{\tau}}^{\downarrow}$ is defined analogously. As discussed in Sec. 2.5.3, there are no canonical requirements on the solution M'. We aim, however, at achieving approximate complementarity: a cell that blocks some users should have a power close to the maximum. In any case, users are only rejected in a cell with $\lambda_i < 1$. We propose the following schemes for determining M':

(a) *Random activation.* Any user requesting service from cell i is admitted at random with a probability of $P(m \in M') = \lambda_i$ for all $m \in M_i$. By linearity of expectation follows

$$\mathbb{E}(\tilde{c}_{ij}^{\downarrow}) = \lambda_i^{\downarrow} c_{ij}^{\downarrow}, \qquad \mathbb{E}(\tilde{c}_{ij}^{\uparrow}) = \lambda_j^{\uparrow} c_{ij}^{\uparrow} \qquad \text{for all } i, j \in \mathcal{N} .$$

We can thus hope to come close to the power values achieved by perfect load control, but this is not guaranteed.

(b) *Random order.* A more sophisticated scheme uses information on the entire distribution of users. In each cell, we deactivate users in a random order until the row sum (in downlink) or the column sum (in uplink) falls below the value for the scaled system. Formally, the constraints on row and column sum are

$$\tilde{\mathbf{C}}^{\downarrow}\mathbf{1} \leq \operatorname{diag}(\boldsymbol{\lambda}^{\downarrow})\,\mathbf{C}^{\downarrow}\mathbf{1}, \qquad \mathbf{1}^t\tilde{\mathbf{C}}^{\uparrow} \leq \mathbf{1}^t\mathbf{C}^{\uparrow}\operatorname{diag}(\boldsymbol{\lambda}^{\uparrow}) .$$

(c) *Row sum/column sum knapsack.* To use more of the cell's resources, we propose to serve as much traffic as possible while fulfilling the constraints of the last scheme:

$$\max_{M' \subset M} \ \mathbf{1}^t \operatorname{diag}(\tilde{\mathbf{C}}^{\uparrow}) \quad \text{s.t.} \ \mathbf{1}^t\tilde{\mathbf{C}}^{\uparrow} \leq \mathbf{1}^t\mathbf{C}^{\uparrow}\operatorname{diag}(\boldsymbol{\lambda}^{\uparrow}) \qquad\qquad \max_{M' \subset M} \ \mathbf{1}^t\tilde{\boldsymbol{\tau}}^{\downarrow} \quad \text{s.t.} \ \tilde{\mathbf{C}}^{\downarrow}\mathbf{1} \leq \operatorname{diag}(\boldsymbol{\lambda}^{\downarrow})\,\mathbf{C}^{\downarrow}\mathbf{1}$$

This problem is called a *knapsack problem* (*Schrijver*, 1986); we solve it with integer programming methods.

(d) *Conservative multiple knapsack.* To guarantee that we obtain a solution fulfilling the soft-capacity constraints, we exploit the monotonicity of power and restrict the produced coupling matrix elementwise:

$$\max_{M' \subset M} \ \mathbf{1}^t \operatorname{diag}(\tilde{\mathbf{C}}^{\uparrow}) \quad \text{s.t.} \ \tilde{\mathbf{C}}^{\uparrow} \leq \mathbf{C}^{\uparrow}\operatorname{diag}(\boldsymbol{\lambda}^{\uparrow}) \qquad\qquad \max_{M' \subset M} \ \mathbf{1}^t\tilde{\boldsymbol{\tau}}^{\downarrow} \quad \text{s.t.} \ \tilde{\mathbf{C}}^{\downarrow} \leq \operatorname{diag}(\boldsymbol{\lambda}^{\downarrow})\,\mathbf{C}^{\downarrow}$$

The maximization problems actually decompose into one multiple knapsack problem per cell; again, we solve it with integer programming.

To find out whether perfect load control is a sensible assumption, we have to compare network performance under discrete load control schemes to the impression of performance that perfect load control gives. If the deviation is large, perfect load control fails to inform us of the real merits and problems of a network design. If the deviation is limited at least for some reasonable admission scheme, this is a hint that perfect load control is a valid simplification.

There are two main approaches for this comparison: worst-case analysis and empirical analysis. In a worst-case analysis, we find parameters for network and snapshot parameters such that the difference is as large as possible. In an empirical analysis, we feed the models with data that is typical for practical cases and observe how large the difference is on average.

We abstain from worst-case analysis, because it exhibits a typical problem: in the worst case, perfect load control is a bad assumption. Instances can be constructed for which the scaling factors do not have any relation to a feasible user-based admission policy. Results of this analysis, however, are not helpful because the worst case may occur infrequently in practice. Consequently, we assess the validity of the perfect load control assumption empirically. Because typical data in practical cases is related to the random distribution on snapshots and expected network performance, this analysis is postponed to Sec. 4.4.1 in the next chapter.

3.3 Generalized pole equations

Pole equations are used in the context of the classical system model for computing cell load and for analyzing the behavior of a single cell. Especially in early research on CDMA systems, a number of results have been obtained by analyzing the stochastics of simplified pole equations. For example, *Sipilä et al.* (2000) calculate the maximum number of users and maximum data rate in downlink; *Veeravalli & Sendonaris* (1999) investigate the stochastic relation between user load and blocking and the trade-off between coverage and capacity in uplink; and *Viterbi & Viterbi* (1993) determine the capacity of a cell for a given maximum blocking ratio.

Pole equations rely on the notion of *other-to-own-cell interference ratio*, a performance indicator defined for any radio receiver in a cellular network with intra-cell interference. It measures the ratio of powers received from senders outside the own cell to that of senders within the cell and is often called *little i* or just i (e. g., *Laiho et al.*, 2002). To avoid confusion with the index i, we will use the symbol ι in the following. The frequency reuse factor (f-factor) is equivalent to other-to-own-cell interference ratio (e. g., *Dehghan et al.*, 2000).

If a given snapshot is served by the network, then the other-to-own-cell interference ratio at cell i is classically defined as

$$\frac{\sum_{j \neq i} \sum_{m \in M_j} \gamma^{\uparrow}_{mj} \alpha^{\uparrow}_m p^{\uparrow}_m}{\sum_{m \in M_i} \gamma^{\uparrow}_{mi} \alpha^{\uparrow}_m p^{\uparrow}_m} = \frac{\sum_{j \neq i} c^{\uparrow}_{ij} \bar{p}^{\uparrow}_j}{c^{\uparrow}_{ii} \bar{p}^{\uparrow}_i} . \tag{3.32}$$

In the present section, we adapt the classic definitions of other-to-own-cell interference ratio to our system model and derive a generalized pole equation as an alternative means of calculating cell powers. This will turn out to be a useful tool for analysis in the following chapters; it furthermore reveals the connection between the interference-coupling systems and the classical pole equations known from engineering literature.

Alternative formulation for an uplink cell. In the uplink, we define the *other-to-own-cell interference ratio* for any solution to (3.25) and any cell $i \in \mathcal{N}$ with $\lambda^{\uparrow}_i c^{\downarrow}_{ii} > 0$ as

$$\iota^{\uparrow}_i := \frac{\sum_{j \neq i} \lambda^{\uparrow}_j c^{\uparrow}_{ij} \bar{p}^{\uparrow}_j}{\lambda^{\uparrow}_i c^{\uparrow}_{ii} \bar{p}^{\uparrow}_i} \tag{3.33}$$

This definition introduces scaling factors into the classical form (3.32). By substituting (3.33) into (3.21), we can rewrite the equations (3.25a) with the coupling between

Figure 3.4: Uplink other-to-own-cell interference ratio vs. served traffic

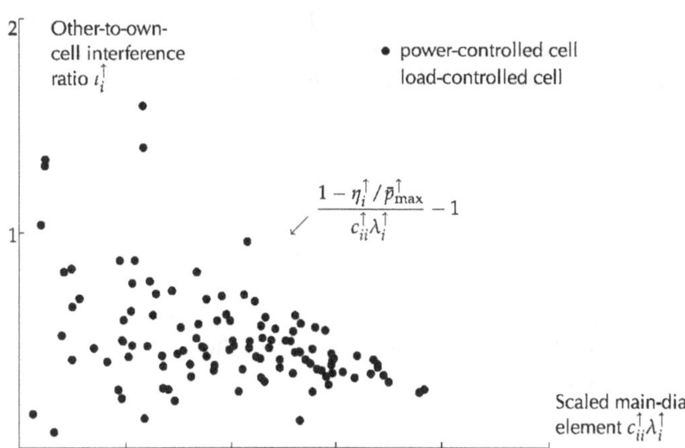

cells singled out into the value $\iota_i^\uparrow$:

$$\bar{p}_i^\uparrow = (1+\iota_i^\uparrow)\lambda_i^\uparrow c_{ii}^\uparrow \bar{p}_i^\uparrow + \eta_i^\uparrow . \tag{3.34}$$

We call $(1+\iota_i^\uparrow)c_{ii}^\uparrow$ the *uplink (cell) load factor* of cell i. Solving (3.34) for $\iota_i^\uparrow$ yields

$$\iota_i^\uparrow = \frac{1-\eta_i^\uparrow/\bar{p}_i^\uparrow}{c_{ii}^\uparrow\lambda_i^\uparrow} - 1 \le \frac{1-\eta_i^\uparrow/\bar{p}_{\max}^\uparrow}{c_{ii}^\uparrow\lambda_i^\uparrow} - 1 .$$

The first identity allows to compute the other-to-own-cell interference ratio of one cell in asymptotic time $O(1)$, whereas the definition (3.33) requires $O(n)$ steps per cell.[2] The bound is attained by load-controlled cells. The values of $\iota^\uparrow$ for cells in an example network are depicted in Fig. 3.4.

By combining the pole equation (3.34) with the complementarity condition (3.23), we obtain an alternative complementarity formulation for the uplink:

Lemma 3.8: *For any solution to the uplink complementarity system (3.23) and for any cell $i \in \mathcal{N}$ with $c_{ii}^\uparrow\lambda_i^\uparrow > 0$, the following identities hold:*

$$\bar{p}_i^\uparrow = \begin{cases} \frac{\eta_i^\uparrow}{1-(1+\iota_i^\uparrow)c_{ii}^\uparrow} & \textit{if } (1+\iota_i^\uparrow)c_{ii}^\uparrow < \frac{\bar{p}_{\max}^\uparrow-\eta_i^\uparrow}{\bar{p}_{\max}^\uparrow}, \\ \bar{p}_{\max}^\uparrow & \textit{otherwise.} \end{cases} \tag{3.35a}$$

$$\lambda_i^\uparrow = \begin{cases} 1 & \textit{if } (1+\iota_i^\uparrow)c_{ii}^\uparrow < \frac{\bar{p}_{\max}^\uparrow-\eta_i^\uparrow}{\bar{p}_{\max}^\uparrow}, \\ \frac{\bar{p}_{\max}^\uparrow-\eta_i^\uparrow}{\bar{p}_{\max}^\uparrow(1+\iota_i^\uparrow)c_{ii}^\uparrow} & \textit{otherwise.} \end{cases} \tag{3.35b}$$

□

We call the value $(\bar{p}_{\max}^\uparrow - \eta_i^\uparrow)/\bar{p}_{\max}^\uparrow$ the *critical (uplink) load factor* of cell i.

[2] Basically, we only need a fixed number of arithmetic operations and save a loop over all indices; a definition of asymptotic running time is given, e.g., by *Cormen, Leiserson, & Rivest* (1990).

Alternative formulation for a downlink cell. For a snapshot served in downlink, the other-to-own-cell interference ratio $\bar{\iota}_m^{\downarrow}$ is defined for each mobile $m \in M$. There is only one sender per cell, which simplifies the definition of the other-to-own-cell interference ratio. We depart from the definition of the *downlink other-to-own-cell interference ratio* for mobile m as

$$\bar{\iota}_m^{\downarrow} := \frac{\sum_{j \neq i} \gamma_{jm}^{\downarrow} \bar{p}_j^{\downarrow}}{\omega_m \gamma_{im}^{\downarrow} \bar{p}_i^{\downarrow}}; \tag{3.36}$$

we assume here that $\omega_m > 0$. Our definition differs from the classical form in including user-specific orthogonality. We discuss the relation to previous pole equations below.

For a cell-wise formulation, an average other-to-own-cell interference ratio is used. There is no formal definition of this quantity in the standard literature. We propose the following weighted average:

$$\frac{\sum_{m \in M_i} \omega_m \ell_m^{\downarrow} \bar{\iota}_m^{\downarrow}}{\sum_{m \in M_i} \omega_m \ell_m^{\downarrow}} = \frac{\sum_{m \in M_i} \ell_m^{\downarrow} \sum_{j \neq i} \frac{\gamma_{jm}^{\downarrow}}{\gamma_{im}^{\downarrow}} \bar{p}_j^{\downarrow}}{\sum_{m \in M_i} \omega_m \ell_m^{\downarrow} \bar{p}_i^{\downarrow}} = \frac{\sum_{j \neq i} c_{ij}^{\downarrow} \bar{p}_j^{\downarrow}}{c_{ii}^{\downarrow} \bar{p}_i^{\downarrow}}.$$

The last form in this transformation can be defined for a solution to the complementarity system (3.26) without referring to a mobile. We thus define the average other-to-own-cell interference ratio for cell i in downlink to be

$$\bar{\iota}_i^{\downarrow} := \frac{\sum_{j \neq i} c_{ij}^{\downarrow} \bar{p}_j^{\downarrow}}{c_{ii}^{\downarrow} \bar{p}_i^{\downarrow}} \tag{3.37}$$

for any cell with $c_{ii}^{\downarrow} > 0$. This facilitates an alternative version of the coupling system:

$$\bar{p}_i^{\downarrow} = (1 + \bar{\iota}_i^{\downarrow}) \lambda_i^{\downarrow} c_{ii}^{\downarrow} \bar{p}_i^{\downarrow} + \lambda_i^{\downarrow} p_i^{(\eta)} + p_i^{(c)}. \tag{3.38}$$

In analogy to the uplink, we call the expression $(1 + \bar{\iota}_i^{\downarrow}) c_{ii}^{\downarrow}$ the *downlink (cell) load factor.* Again, we can derive an identity for $\bar{\iota}_i^{\downarrow}$:

$$\bar{\iota}_i^{\downarrow} = \frac{1 - (\lambda_i^{\downarrow} p_i^{(\eta)} + p_i^{(c)}) / \bar{p}_i^{\downarrow}}{\lambda_i^{\downarrow} c_{ii}^{\downarrow}} - 1. \tag{3.39}$$

The other-to-own-cell interference ratio function for a test mobile is plotted for an example network in Fig. 3.5. (The concept of a test mobile is explained in Sec. 3.4.2.) The image shows that the values are highest at cell borders and low at cell centers.

The downlink counterpart to Lemma 3.8 is derived by applying (3.24) to (3.38):

Lemma 3.9: *For any solution to the downlink complementarity system* (3.24) *and for any cell*

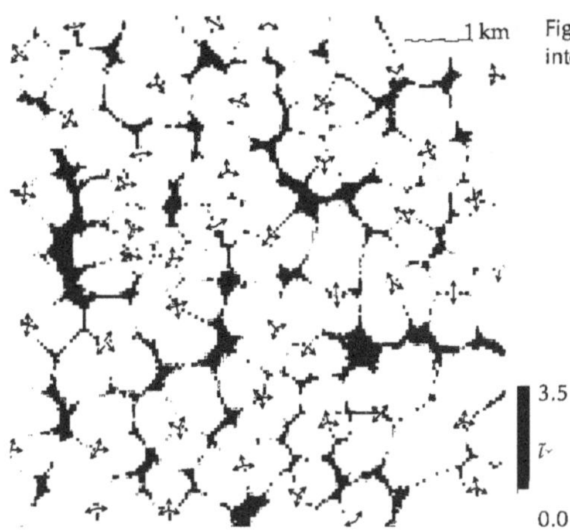

Figure 3.5: Other-to-own-cell interference ratio (downlink)

$i \in \mathcal{N}$ with $c_{ii}^{\downarrow} > 0$, the following identities hold:

$$\bar{p}_i^{\downarrow} = \begin{cases} \frac{p_i^{(c)} + p_i^{(\eta)}}{1-(1+\bar{\iota}_i^{\downarrow})c_{ii}^{\downarrow}} & \text{if } (1+\bar{\iota}_i^{\downarrow})c_{ii}^{\downarrow} < \frac{\bar{p}_{\max}^{\downarrow} - p_i^{(c)} - p_i^{(\eta)}}{\bar{p}_{\max}^{\downarrow}}, \\ \bar{p}_{\max}^{\downarrow} & \text{otherwise.} \end{cases} \tag{3.40a}$$

$$\lambda_i^{\downarrow} = \begin{cases} 1 & \text{if } (1+\bar{\iota}_i^{\downarrow})c_{ii}^{\downarrow} < \frac{\bar{p}_{\max}^{\downarrow} - p_i^{(c)} - p_i^{(\eta)}}{\bar{p}_{\max}^{\downarrow}}, \\ \frac{\bar{p}_{\max}^{\downarrow} - p_i^{(c)}}{\bar{p}_{\max}^{\downarrow}(1+\bar{\iota}_i^{\downarrow})c_{ii}^{\downarrow} + p_i^{(\eta)}} & \text{otherwise.} \end{cases} \tag{3.40b}$$

□

We call the value $(\bar{p}_{\max}^{\downarrow} - p_i^{(c)} - p_i^{(\eta)}) / \bar{p}_{\max}^{\downarrow}$ the *critical (downlink) load factor* of cell i.

Relation to classical pole equations. Examples for the traditional use of pole equations can be found in the books by *Holma & Toskala* (2001), *Laiho et al.* (2002), and *Nawrocki et al.* (2006, Ch. 6.3). In the uplink, they are usually identical to our form (e. g., *Veeravalli & Sendonaris*, 1999). In the downlink, however, we deviate from the classical formulation (e. g., *Sipilä et al.*, 2000; *Viterbi & Viterbi*, 1993) because our notion (3.36) of a mobile's other-to-own-cell interference ratio includes user-specific orthogonality. The classical model assumes a global orthogonality value. It furthermore uses the *average path loss*, which is hidden in our definition (3.12) of the downlink noise load $p_i^{(\eta)}$. If we assume that the noise $\eta_m^{\downarrow}$ is a constant value $\eta^{\downarrow}$ for all mobiles and define the average path loss in cell i as

$$1/\bar{\gamma}_i^{\downarrow} := \textstyle\sum_m \frac{\ell_m^{\downarrow}}{\gamma_{im}^{\downarrow}} / \sum_m \ell_m^{\downarrow},$$

i. e., weighted by the user load factors, then it holds that

$$p_i^{(\eta)} = \textstyle\sum_{m \in M_i} \frac{\eta_m^{\downarrow}}{\gamma_{im}^{\downarrow}} \ell_m^{\downarrow} = \frac{\eta^{\downarrow}}{\bar{\gamma}_i^{\downarrow}} \sum_{m \in M_i} \ell_m^{\downarrow}.$$

Table 3.1: Overview of performance indicators

	performance indicator	symbol
user-based	E_c covered area	$\|A^{(E_c)}\|$
	E_c/I_0 covered area	$\|A^{(E_c/I_0)}\|$
	covered area	$\|A^{(C)}\|$
	soft handover area	$\|A^{(SHO)}\|$
	pilot-polluted area	$\|A^{(PP)}\|$
uplink	cell load	$\bar{L}^{\uparrow}$
	grade of service	$\bar{\lambda}^{\uparrow}$
	other-to-own-cell interference	$\bar{\iota}^{\uparrow}$
downlink	cell load	$\bar{L}^{\downarrow}$
	grade of service	$\bar{\lambda}^{\downarrow}$
	other-to-own-cell interference	$\bar{\iota}^{\downarrow}$
	total transmit power	$\sum_i \bar{p}_i^{\downarrow}$

The average path loss is used in this fashion in the cited sources.

The classical pole equations were only valid for a regular cell structure and uniformly distributed traffic; our generalizations extend them to instances with arbitrary data. If adapted as above, all previous results apply in the general case. Besides, we can gain insights and intuition on the behavior of the complementarity systems through the generalized pole equations; an example is given in Sec. 4.2.3.

3.4 Performance indicators

We now define indicators for the performance of a UMTS radio network based on a solution to the complementarity system. Our metrics are basically equivalent to those in commercial planning software (e. g., *AWE Communications GmbH*, 2005*a*; *Cosiro GmbH*, 2006; *Lustig et al.*, 2004). These tools, however, usually feature service-specific analysis, while we consider only normalized, aggregated traffic. Furthermore, we are restricted to the performance that can be evaluated in static models; as discussed in Sec. 2.5, this notably excludes the performance of data services. We define performance indicators at *cell level* and at *user level*. A similar distinction between sector-based statistics and geographically-based performance is made in software tools (e. g., *Schema Ltd*, 2003*b*). All performance indicators are summarized in Tab. 3.1; example values for realistic network configurations are contained in Tabs. 5.4 and B.7.

3.4.1 Cell-based performance indicators

For each cell-based performance indicator, we define a value per cell and a network aggregate.

Cell load. A cell's *load* is a fractional value that specifies to what extent the radio resources are consumed; it takes values in the interval $[0, 1]$. Our definitions match the standard ones (e. g., *Holma & Toskala*, 2001, Ch. 9.4).

A cell's *uplink load* $L^{\uparrow}$ is measured with respect to the empty system in terms of the noise rise defined in (2.5). For cell $i \in \mathcal{N}$ it is defined as

$$L_i^{\uparrow} := 1 - 1/\mathrm{NR}_i^{\uparrow} = 1 - \eta_i^{\uparrow}/\bar{p}_i^{\uparrow}\,. \tag{3.41}$$

With this definition, a cell i in an empty system ($p_i^{\uparrow} = \eta_i^{\uparrow}$) has load $0\,\%$. A value of $L_i^{\uparrow} = 100\,\%$ cannot be attained; it corresponds to an infinite noise rise. The average load in the network is determined by the *average noise rise*

$$\overline{\mathrm{NR}}^{\uparrow} := \textstyle\sum_i \mathrm{NR}_i^{\uparrow}/n \tag{3.42}$$

as

$$\bar{L}^{\uparrow} := 1 - 1/\overline{\mathrm{NR}}^{\uparrow}\,. \tag{3.43}$$

A cell's downlink load is defined with respect to the full cell as the percentage of total available power $p_{\mathrm{max}}^{\downarrow}$ needed for serving the present users:

$$L_i^{\downarrow} := \bar{p}_i^{\downarrow}/p_{\mathrm{max}}^{\downarrow}\,. \tag{3.44}$$

Because $0 < p_i^{(\mathrm{c})} \le \bar{p}_i^{\downarrow} \le \bar{p}_{\mathrm{max}}^{\downarrow} < p_{\mathrm{max}}^{\downarrow}$, the downlink cell load does not take on the extreme values $0\,\%$ or $100\,\%$. Its average is

$$\bar{L}^{\downarrow} := \textstyle\sum_i \bar{p}_i^{\downarrow} / \sum_i p_{\mathrm{max}}^{\downarrow}\,. \tag{3.45}$$

In addition, we use the *total transmitted power* $\sum_i \bar{p}_i^{\downarrow}$ as a performance indicator.

Grade of service. The grade of service of a cell i traditionally is the ratio of the number of admitted users over the total number of users requesting access. With the new system model, the grade of service of cell i in uplink and downlink is identical to the traffic scaling value $\lambda_i^{\uparrow}$ and $\lambda_i^{\downarrow}$. It is aggregated for all cells according to the total offered traffic:

$$\bar{\lambda}^{\downarrow} := \textstyle\sum_i \tau_i^{\downarrow}\lambda_i^{\downarrow} / \sum_i \tau_i^{\downarrow}\,, \tag{3.46a}$$

$$\bar{\lambda}^{\uparrow} := \textstyle\sum_i c_{ii}^{\uparrow}\lambda_i^{\uparrow} / \sum_i c_{ii}^{\uparrow}\,. \tag{3.46b}$$

The load and the grade of service per cell of a radio network in Berlin are illustrated in Fig. 3.6.

Other-to-own-cell interference ratio. Intra-cell interference is inevitable in W-CDMA, but inter-cell interference depends on the network design and can be reduced. The other-to-own-cell interference ratio is therefore informative for network planning. Our

Figure 3.6: Cell performance in a radio network in Berlin, downlink

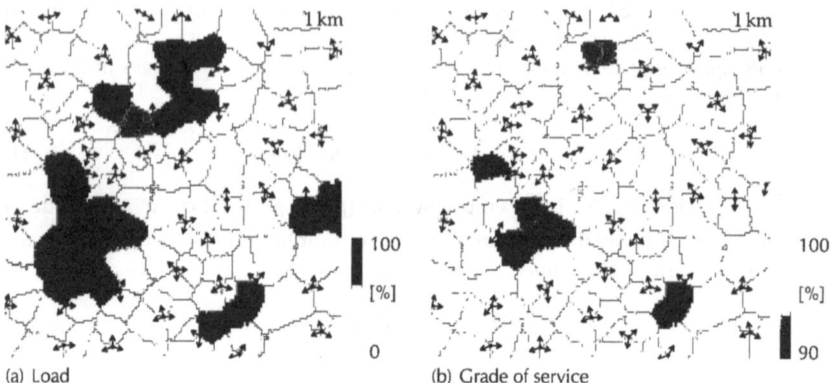

(a) Load (b) Grade of service

definitions (3.33) and (3.37) modify the classical ones (*Laiho et al.*, 2002, Ch. 2.5.1.10) as discussed on p. 55.

For the entire network, the other-to-own-cell interference ratio is weighted by the served traffic and averaged:

$$\bar{\iota}^{\uparrow} := \textstyle\sum_i \lambda_i^{\uparrow} c_{ii}^{\uparrow} \iota_i^{\uparrow} \,/\, \textstyle\sum_i \lambda_i^{\uparrow} c_{ii}^{\uparrow}, \tag{3.47a}$$

$$\bar{\iota}^{\downarrow} := \textstyle\sum_i \lambda_i^{\downarrow} c_{ii}^{\downarrow} \bar{\iota}_i^{\downarrow} \,/\, \textstyle\sum_i \lambda_i^{\downarrow} c_{ii}^{\downarrow}. \tag{3.47b}$$

The denominators can only vanish if no traffic at all is served; this case is excluded.

Note that for our definition in downlink, the values from literature have to be adapted. For example, the reference value of 65 % stated by *Nawrocki et al.* (2006, Ch. 10.2.4.6) for the case of three-sectorized Macro-cells, which applies to all cases considered in this thesis, has to be scaled with the average orthogonality-loss factor. For $\omega \equiv 0.6$, the adapted value is $\bar{\iota}^{\downarrow} = 108.0\,\% = 65\,\%/0.6$.

3.4.2 User-based performance indicators

User-based performance indicators reflect under which circumstances a user can access a given service at a given location. We have discussed above that we refrain from service-specific analysis; for the sake of simplicity, we also limit our analysis to performance indicators that are based on checking a logical condition per user. For example, this includes coverage, but it excludes the mobile transmit power needed at a given location.

Extension to planning area and aggregation. To gain more information from evaluating a single snapshot, we extend the definition of user-based performance figures from the users in a snapshot to the total planning area. To this end, we hypothetically place a *test mobile* at any location in the area. The test mobile does not belong to the snapshot and has no effect on the network's performance. In this way, a single solution of the system model suffices to determine the part of the planning area, for which the condition is met.

For example, we determine the set $A^{(\mathrm{E_c/I_0})}$ of points in which a test mobile has $\mathrm{E_c/I_0}$ coverage by checking condition (2.8c). This set is represented by an indicator function $\mathbf{1}_{\mathrm{E_c/I_0}} : A \to \{0,1\}$ taking the value 1 wherever the condition holds and 0 elsewhere. The test mobile approach makes a crucial difference here, because an additional real mobile would change the cell power levels.

We use the fraction $|A^{(\mathrm{E_c/I_0})}|/|A|$ of the area that is $\mathrm{E_c/I_0}$-covered as a representative aggregate. To take into account that some areas are more frequented than others, we define additional versions of the user-based performance indicators weighted with the total normalized user load intensity function $T^{\downarrow}_{\ell} : A \to \mathbb{R}_{+}$. This function indicates how much aggregated traffic is likely to occur in a given area; it is precisely defined in Sec. 4.2 below. In analogy to the calculation of the covered area as

$$|A^{(\mathrm{E_c/I_0})}| = \int_A \mathbf{1}_{\mathrm{E_c/I_0}}(x)\mathrm{d}x\,, \tag{3.48a}$$

we define the weighted version of the coverage measure as

$$|A^{(\mathrm{E_c/I_0})}|^{\downarrow}_{\ell} := \int_A \mathbf{1}_{\mathrm{E_c/I_0}}(x)\, T^{\downarrow}_{\ell}(x)\mathrm{d}x\,. \tag{3.48b}$$

The construction applies to all user-based performance measures below and also to a three-dimensional scenario. In the following, we will only define the respective sets; analogous unweighted and weighted versions of the performance indicators are understood.

For the sake of brevity, we will also denote the fraction of the area A that is $\mathrm{E_c/I_0}$ covered by $|A^{(\mathrm{E_c/I_0})}|$ instead of $|A^{(\mathrm{E_c/I_0})}|/|A|$. The accompanying unit % will clarify that the value is to be understood as percentage of the total area.

Coverage. Recall the definition of CPICH coverage given in Sec. 2.4.2. The $\mathrm{E_c}$-covered points in the planning area are those fulfilling condition (2.6):

$$A^{(\mathrm{E_c})} := \{x \in A \,|\, \max_i \gamma^{\downarrow}_{ix} p^{(\mathrm{P})}_i \geq \pi_{\mathrm{E_c}}\}\,. \tag{3.49}$$

The signal level and $\mathrm{E_c}$ coverage of a network in Lisbon are depicted in Fig. 2.9(a). The $\mathrm{E_c/I_0}$-covered subset of A is defined with condition (2.8c) as

$$A^{(\mathrm{E_c/I_0})} := \left\{x \in A \,|\, \max_i \frac{\gamma^{\downarrow}_{ix} p^{(\mathrm{P})}_i}{\eta^{\downarrow}_m + \sum_{j \in \mathcal{N}} \gamma^{\downarrow}_{jx} \bar{p}^{\downarrow}_j} \geq \mu_{\mathrm{E_c/I_0}}\right\}. \tag{3.50}$$

The quality of the pilot signal and the area $A^{(\mathrm{E_c/I_0})}$of $\mathrm{E_c/I_0}$-coverage are illustrated in Fig. 2.9(b). We define the set of *covered* points as the intersection of the two sets:

$$A^{(\mathrm{C})} := A^{(\mathrm{E_c})} \cap A^{(\mathrm{E_c/I_0})}\,. \tag{3.51}$$

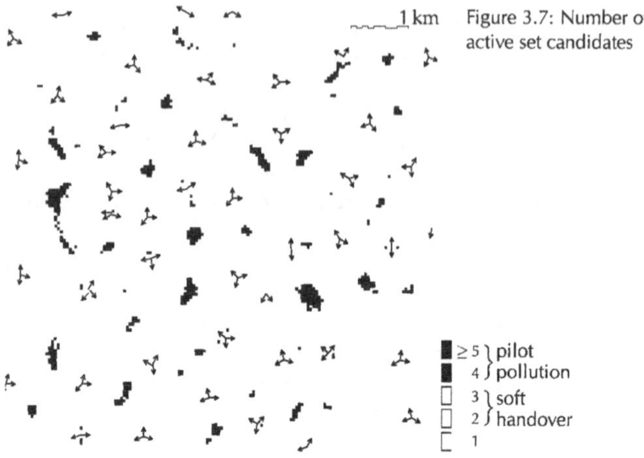

Figure 3.7: Number of active set candidates

Soft handover probability and pilot pollution. Soft handover is not considered in our system model, but we define performance indicators that allow conclusions on the network behavior under soft handover. Each mobile keeps track of an *active set* of base stations whose pilot signal is received at most a factor of $\theta^{(\text{SHO})} > 0$ (the soft handover margin) weaker than the strongest pilot signal received. We define the number of candidates for the active set at point $x \in A$ as

$$n^{(\text{SHO})}(x) := \left|\{i \in \mathcal{N} \,|\, \max_j \gamma^{\downarrow}_{jx} p^{(\text{P})}_j / \theta^{(\text{SHO})} \leq \gamma^{\downarrow}_{ix} p^{(\text{P})}_i \}\right| .$$

The function $n^{(\text{SHO})}$ is shown for the Berlin scenario in Fig. 3.7. It is desirable to have more than one cell in the active set for mobiles further away from their server, but the number of pilot signals that can be monitored is limited. We denote the maximum active set size by $n^{(\text{SHO})}_{\max}$. If more than $n^{(\text{SHO})}_{\max}$ servers are candidates for the active set, the mobile is not able to consistently monitor all of them. This phenomenon is called *pilot pollution*. The *soft handover area* and the *pilot pollution area* are defined as

$$A^{(\text{SHO})} := \{x \in A \,|\, 2 \leq n^{(\text{SHO})}(x) \leq n^{(\text{SHO})}_{\max}\} \cap A^{(\text{C})} , \tag{3.52}$$

$$A^{(\text{PP})} := \{x \in A \,|\, n^{(\text{SHO})}(x) > n^{(\text{SHO})}_{\max}\} \cap A^{(\text{C})} . \tag{3.53}$$

In general, pilot pollution should be as small as possible; a soft handover area of about 30 % is recommended (*Holma & Toskala*, 2001).

Things to remember: Interference-coupling complementarity systems

- For a given snapshot, the CIR-equations are condensed into the *interference-coupling equation systems*, which model power control, but neither outage nor load control:

$$\bar{p}^{\uparrow} = C^{\uparrow}\bar{p}^{\uparrow} + \eta^{\uparrow} \qquad\qquad \bar{p}^{\downarrow} = C^{\downarrow}\bar{p}^{\downarrow} + p^{(\eta)} + p^{(c)}$$

These equation systems allow to calculate cell powers without explicitly considering users. The dimension is the number of cells in the network. The parameters are:

$\bar{p}$	cell powers	$p^{(\eta)}$	downlink noise load
C	interference-coupling matrix	$p^{(c)}$	power on DL common channels
$\eta^{\uparrow}$	noise at antenna		

- We add the new assumption of *perfect load control:*
  - Load controlled cells command all their users to continuously scale their load factor by a common value represented by the cell's *scaling factor* $\lambda \in [0,1]$.
  - A load-controlled cell uses all its power to serve the maximum of user demand. This entails *complementarity relations* between power and scaling values.

- The new resulting model of interference-coupling complementarity systems extend the coupling equation to soft blocking at cell level:

$$\bar{p}^{\uparrow} = C^{\uparrow}\operatorname{diag}(\lambda^{\uparrow})\,\bar{p}^{\uparrow} + \eta^{\uparrow} \qquad\qquad \bar{p}^{\downarrow} = \operatorname{diag}(\lambda^{\downarrow})\,(C^{\downarrow}\bar{p}^{\downarrow} + p^{(\eta)}) + p^{(c)}$$

$$0 = (1-\lambda^{\uparrow})^{t}(\bar{p}^{\uparrow}_{\max} - \bar{p}^{\uparrow}) \qquad\qquad 0 = (1-\lambda^{\downarrow})^{t}(\bar{p}^{\downarrow}_{\max} - \bar{p}^{\downarrow})$$

We specify efficient iterative schemes to solve these two models.

- With the suitable redefinition of the other-to-own-cell interference ratio ι we derive generalized *pole equations*

$$\bar{p}^{\uparrow}_i = \begin{cases} \frac{\eta^{\uparrow}_i}{1-(1+\iota^{\uparrow}_i)c^{\uparrow}_{ii}} & \text{if } (1+\iota^{\uparrow}_i)c^{\uparrow}_{ii} < \frac{\bar{p}^{\uparrow}_{\max}-\eta^{\uparrow}_i}{\bar{p}^{\uparrow}_{\max}} \\ \bar{p}^{\uparrow}_{\max} & \text{otherwise} \end{cases} \qquad \bar{p}^{\downarrow}_i = \begin{cases} \frac{p^{(c)}_i+p^{(\eta)}_i}{1-(1+\bar{\iota}^{\downarrow}_i)c^{\downarrow}_{ii}} & \text{if } (1+\bar{\iota}^{\downarrow}_i)c^{\downarrow}_{ii} < \frac{\bar{p}^{\downarrow}_{\max}-p^{(c)}_i-p^{(\eta)}_i}{\bar{p}^{\downarrow}_{\max}} \\ \bar{p}^{\downarrow}_{\max} & \text{otherwise} \end{cases}$$

$$\lambda^{\uparrow}_i = \begin{cases} 1 & \text{if } (1+\iota^{\uparrow}_i)c^{\uparrow}_{ii} < \frac{\bar{p}^{\uparrow}_{\max}-\eta^{\uparrow}_i}{\bar{p}^{\uparrow}_{\max}} \\ \frac{\bar{p}^{\uparrow}_{\max}-\eta^{\uparrow}_i}{\bar{p}^{\uparrow}_{\max}(1+\iota^{\uparrow}_i)c^{\uparrow}_{ii}} & \text{otherwise} \end{cases} \qquad \lambda^{\downarrow}_i = \begin{cases} 1 & \text{if } (1+\bar{\iota}^{\downarrow}_i)c^{\downarrow}_{ii} < \frac{\bar{p}^{\downarrow}_{\max}-p^{(c)}_i-p^{(\eta)}_i}{\bar{p}^{\downarrow}_{\max}} \\ \frac{\bar{p}^{\downarrow}_{\max}-p^{(c)}_i}{\bar{p}^{\downarrow}_{\max}(1+\bar{\iota}^{\downarrow}_i)c^{\downarrow}_{ii}+p^{(\eta)}_i} & \text{otherwise} \end{cases}$$

describing cell power and grade of service as a function of the *load factor* $(1+\iota_i)c_{ii}$; the case distinction reflects the *critical load factor.*

- We define a set of *performance indicators* including coverage, cell load, and user blocking. All aspects of network performance are thus described at cell level as a function of the coupling matrix.

Expected-interference-coupling estimates for network performance

Radio network planning aims at improving the expected network performance, so we are not interested in network performance on a single snapshot, but on the expected performance for random snapshots. The coupling matrices thus have to be considered random variables subject to a probability distribution induced by the distribution on snapshots, and we are interested in the stochastics of the performance indicators.

Simulation methods are commonly used for determining mean values of performance indicators, but they are inherently too time consuming for use in heavy-duty optimization tools; therefore, faster approaches are needed. While Monte Carlo methods can yield an arbitrary accuracy if sufficient time is granted, high (absolute) precision is dispensable for taking intermediate planning decisions. For a successful optimization campaign, the ability to quickly discriminate between design alternatives is paramount. The right decision can be made in short time, if accuracy is sacrificed in a controlled fashion. The practical relevance of quick estimation techniques is apparent from the fact that many commercial software tools advertise fast proprietary evaluation methods besides Monte Carlo simulation (*Aircom International ltd*, 2007; *Cosiro GmbH*, 2006; *Ericsson AB*, 2006; *Lustig et al.*, 2004).

In this chapter, we develop methods for estimating the expected network performance with little computational effort. The basic idea is to calculate approximations to the mean values of capacity-related performance indicators based on the mean coupling matrix. The scheme depends on suitable choices of the performance model and of the random model. With the interference coupling complementarity systems, we have a detailed model that reflects the relations between cells. We restrict the random model on snapshots to exclude shadow fading, calculating with the medians of attenuation (the deterministic path loss component) instead. The resulting method of *expected interference coupling with medians of attenuation* is tailored to the common representation of planning data in computer software. We complement it with a specialized method that calculates better estimates of the grade of service using second-order moments.

Besides the method itself, this chapter contributes the thorough analysis of the expected coupling method and its validation as a suitable tool for network planning. Our investigations comprise analytical and empirical studies. On the analytical side, we use the new generalized pole equations in a simplified setting; the results explain how the service mix determines the variance of the coupling matrix and thereby the quality of the estimates. In our computational studies, we essentially demonstrate that the method is sufficiently informative for typical applications in network planning.

The remainder of this chapter is structured as follows: We introduce the accurate reference method of Monte Carlo simulation in Sec. 4.1. We define the expected coupling estimates and analyze their accuracy in Sec. 4.2. Refined estimates for the grade of service are developed in Sec. 4.3. In Sec. 4.4, we conduct extensive computational experiments to analyze the accuracy of our new estimates and assess the validity of perfect load control in realistic settings. We draw conclusions on network modeling and performance evaluation in Sec. 4.5.

Related work. Random quantities are often represented by their mean in a first-order approximation; in so far the expected coupling approach is a canonical choice. *Dziong et al.* (1999) seem to be the first to propose the technique for CDMA radio networks; we reproduce some of their results in our below analysis. Similar schemes have been used occasionally in scientific works on network planning and optimization (*Sobczyk*, 2005; *Türke & Koonert*, 2005) and also in commercial products (*Buddendick*, 2005). None of the cited works, however, conduct a thorough analysis.

The "static method" for performance evaluation by *Türke* (2006) is equivalent to our scheme; his more refined "statistical methods" aim at including shadow fading. Similar approaches for avoiding simulation are developed in the string of work by *Mäder & Staehle* (2004); *Staehle* (2004); *Staehle et al.* (2002, 2003). Whenever shadow fading is considered, however, numerical integration in several dimensions is required.

In a classical paper on CDMA capacity, *Viterbi & Viterbi* (1993) determine the capacity and blocking probability in a single CDMA cell with fixed interference coupling. Our refined estimation scheme can be seen as a generalization.

4.1 The reference method: simplified Monte Carlo simulation

Expected values of performance indicators can be accurately computed with Monte Carlo simulation for an arbitrary random distribution of snapshots. The general principle is introduced in Sec. 2.5.1; its application to our system model is detailed in Algorithm 1 below. The algorithm computes the empirical means $(\tilde{p}, \tilde{\lambda})$ of cell powers and scaling factors; we list it in a generic form that applies to uplink and downlink. In each iteration, the steps listed in lines 2 through 5 are repeated. First, a snapshot is drawn in line 2. In line 3, the interference-coupling matrix is calculated based on the information in the traffic snapshot. Effectively, lines 2 and 3 produce a sample of an interference-coupling matrix from the random distribution on matrices induced by the distribution on snapshots. The complementarity system for the generated matrix is solved in line 4, and the resulting performance indicators are aggregated in line 5.

Algorithm 1 Simplified Monte Carlo simulation

1: **repeat**
2: $(\mu, \alpha, \omega, \tilde{\gamma}) \leftarrow$ DRAWSNAPSHOT ▷ Sec. 2.5.2
3: $C \leftarrow$ COUPLINGMATRIX$(\mu,\alpha,\omega,\tilde{\gamma})$ ▷ Sec. 3.1
4: $(\bar{p}, \lambda) \leftarrow$ SOLVECOMPLEMENTARITY(C) ▷ Sec. 3.2.3
5: $(\tilde{p}, \tilde{\lambda}) \leftarrow$ AGGREGATEPERFORMANCE$(\bar{p},\lambda)$ ▷ Sec. 3.4
6: **until** $P(\mathbb{E}(\log \tilde{p}) \in [\log \tilde{p} \pm 1\,\%]) \geq 0.99$

Algorithm 1 is basically equivalent to the simulation routines implemented in commercial tools up to the omission of service specific analysis discussed in Sec. 3.4.

Convergence and computation times. In line 6 of the algorithm, we test convergence with confidence intervals (see, for example, *Jain*, 1991) for the mean cell power values. Based on the common assumption that the transmission power of a cell follows a lognormal distribution (*Viterbi & Viterbi*, 1993), the confidence intervals are computed in logarithmic scale. The performance measures are considered to have sufficiently converged once the logarithmic relative error of the empiric mean is at most 1 % with a probability of at least 99 %.

The number of iterations needed to obtain convergence depends on the desired accuracy and on the variance of the cell powers. By applying the law of large numbers, it can be shown that the empirical estimates converge with a rate of $O(\mathrm{Var}(\bar{p})/\sqrt{N})$, where N is the number of samples and hence iterations of the outer loop (*Fishman & Fishman*, 1996; *Madras*, 2002). For double accuracy, the number of iterations therefore has to be quadrupled. No closed formula describing the variance of the cell powers through the input distributions is known; *Staehle* (2004) develops approximations for the variance of transmit powers, but his estimates disregard soft capacity and are conditioned to the case that finite power suffices. As a rule of thumb, a higher variance of shadow fading and a higher percentage of demanding data users increases the simulation time.

In practice, accurate simulation requires the evaluation of thousands of snapshots and a computation time on the order of a few hours. The precise values for the most demanding simulation cases from our computations presented in Sec. 4.4.2 are listed in Tab. 4.1. In all cases, around 13 000 iterations are needed; this took between three and five hours of computation time on high-end PCS, typical equipment for network planning. The iteration numbers roughly match the 10 000 snapshots recommended in an official 3GPP standard (3GPP TSG RAN, 2006*a*) for voice users only. Even ten times more snapshots are recommended for data users in the standard, but this was not necessary with our convergence criterion. The merely 100–500 samples that *Lustig et al.* (2004) deem typical seem to relate to a less accurate evaluation. The time measurements provide an indication of the typical computational effort, but they are not

Table 4.1: Computational effort of Monte Carlo simulation
(medium traffic intensity, complete service mix)

Scenario Network	Berlin reg.	Berlin opt. 1	Lisbon reg.	Lisbon opt.
no. of iterations N	13 850	12 850	12 950	12 800
simulation time [h:mm]	4:21[a]	5:00[a]	3:19[b]	3:06[b]
time for computing estimates [s]	2[a]	3[a]	9[b]	11[b]

[a] Intel™ Pentium™ D CPU, 3.6 GHZ, 3.8 GB RAM
[b] Dual Core AMD Opteron™ 875 CPU, 2.2 GHZ, 32.5 GB RAM

directly comparable to practice. On the one hand, a discrete load-control routine consumes more time per snapshot than our simplified method, so our algorithm should be faster; on the other hand, we did not put specific emphasis on streamlining our implementation, so the computation times listed in the table may be easily reduced.

Evaluation of stochastics of performance indicators. If snapshots vary randomly, the performance indicators defined in Sec. 3.4 are random variables; their stochastics determine expected network performance. We shall briefly describe the nature of the resulting random variables and the aspects on which we focus in the following. Cell-based and user-based performance indicators are treated differently.

The cell-based performance indicators defined in Sec. 3.4.1 are real-valued random vectors. An exemplary vector of mean uplink cell loads for a radio network is depicted in Fig. 4.1(a); the error bars indicate the standard deviation. In principle, all properties of a random variable can be studied; for our purposes, we focus on the expected value. Higher order moments, especially variance, may occasionally be relevant for performance evaluation and optimization, but they are beyond the scope of this thesis.

Among the user-based performance indicators defined in Sec. 3.4.2, we focus on E_c/I_0 coverage, because it is the only one linked to load and interference coupling. The remaining user-based values are E_c coverage, soft handover, and pilot pollution. Their stochastics depend on the model of shadow fading alone. For common shadowing models, they are analyzed considering simple functions of multivariate distributions; this falls out of the primary scope of this thesis. For example, *Kürner* (1998) investigates similar questions using a Markov chain model with special regard to the handover process.

The main indicator for E_c/I_0 coverage is the mean covered area $\mathbb{E}(|A^{(E_c/I_0)}|)$ and its weighted variant $\mathbb{E}(|A^{(E_c/I_0)}|_{\ell}^{\downarrow})$. Under random variation of the snapshots, the indicator function $\mathbf{1}_{E_c/I_0}$ for E_c/I_0 coverage defines a Bernoulli-distributed random variable for each point in the area. For a given point $x \in A$, the random variable $\mathbf{1}_{E_c/I_0}(x)$ evaluates to 1 if the point is covered for a given snapshot and to 0 otherwise.

The expected values of aggregates indicate *if* there is a problem in the network; the level sets of the coverage probability function indicate *where* improvements are needed. The coverage probability function $\mathbb{E}(\mathbf{1}_{E_c/I_0}) : A \to [0,1]$ is depicted for an

Figure 4.1: Stochastic of performance indicators

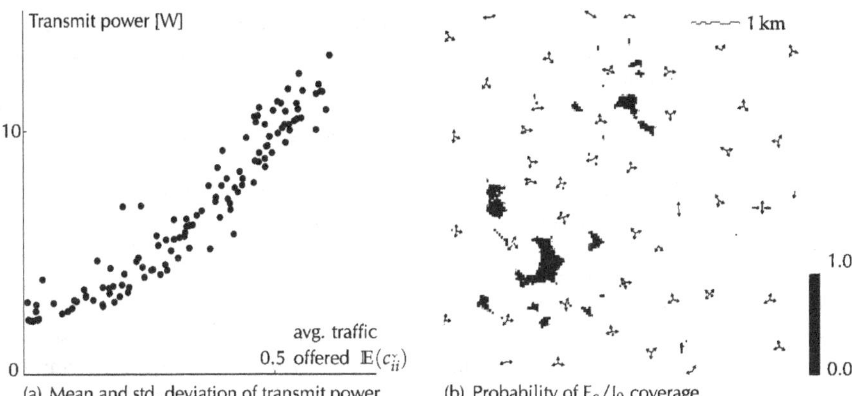

(a) Mean and std. deviation of transmit power (b) Probability of E_c/I_0 coverage

example network in Fig. 4.1(b). Operators usually want to limit outage probabilities to a small value wherever possible. A typical threshold for an acceptable outage probability is 2 %; to assess where this is violated, we are interested in the set $\{x \in A \mid \mathbb{E}(\mathbf{1}_{E_c/I_0}(x)) < 0.98\}$.

4.2 Expected interference coupling with medians of attenuation

In the present section, we propose a method for avoiding simulation and short-cutting the evaluation of UMTS radio networks. The basic idea is to ignore the higher-order moments of the random distribution on the coupling matrix induced by the distribution on snapshots and calculate performance estimates using the expected coupling matrix. This approach introduces inaccuracies, but it requires only a single evaluation of a complementarity system.

Computing the expected coupling matrix, however, is difficult in itself unless the random model for snapshots is restricted. We have to make two assumptions. First, shadow fading does not occur. Second, mobiles always try to connect to the best server. If the shadow fading component follows the usual lognormal distribution, the path loss component of channel gain is the median (not the mean) of the distribution of the gain factors. Our method is therefore called *expected interference coupling with medians of attenuation.*

We will first specify how to calculate the expected interference coupling matrix in our setting in Sec. 4.2.1. The details on how to compute estimates for the expected performance indicators are given in Sec. 4.2.2. We analyze the estimation errors for cell power and grade of service in Sec. 4.2.3.

4.2.1 Calculating the expected interference coupling matrix

We assume that each mobile tries to associate to the cell providing the strongest pilot signal, its *best server.* If shadowing is disregarded, the cell assignments then depend only on the *cell area* defined as

$$A_i := \{x \in A \mid \gamma_i^{\downarrow}(x)p_i^{(\mathrm{P})} \geq \gamma_j^{\downarrow}(x)p_j^{(\mathrm{P})} \text{ for all } j \in \mathcal{N}\} \cap A^{(\mathrm{E_c})}. \tag{4.1}$$

for any $i \in \mathcal{N}$. Recall that $A^{(\mathrm{E_c})}$ has been defined in (3.49) as the set of $\mathrm{E_c}$-covered pixels. More specifically, we say that the *best-server assumption* is fulfilled for a pair of a network and an $\mathrm{E_c}$-covered snapshot, if the sets $\{A_i\}_{i\in\mathcal{N}}$ are measurable and form a partition of A up to a null set, and each mobile is assigned to the cell in whose cell area it is located. In practice, the cell areas may actually overlap, because the path loss functions are specified as rounded values over a discretization of A. We assume that ties are broken in a well-defined manner in this case. Cell areas are, for example, shown in Fig. 3.6.

Recall Def. 4 on p. 30 describing the restricted standard random distribution on snapshots; we use the notation introduced there. For a service r, let

$$\ell_r^{\uparrow} := \mu_r^{\uparrow}\alpha_r^{\uparrow}/(1+\mu_r^{\uparrow}\alpha_r^{\uparrow}), \qquad \ell_r^{\downarrow}(x) := \mu_r^{\downarrow}\alpha_r^{\downarrow}/(1+\omega(x)\mu_r^{\downarrow}\alpha_r^{\downarrow}), \quad x \in A.$$

We define the *total normalized user load intensities* in uplink and downlink as the functions $T_\ell^{\uparrow}, T_\ell^{\downarrow} : A \to \mathbb{R}_+$ with

$$T_\ell^{\uparrow}(x) := \textstyle\sum_r \ell_r^{\uparrow} T_r(x), \qquad T_\ell^{\downarrow}(x) := \sum_r \ell_r^{\downarrow}(x) T_r(x), \quad x \in A. \tag{4.2}$$

The function aggregates the user intensity over all services, weighted by the service's user load; Fig. 5.6(b) depicts an example for the normalized downlink traffic.

If shadow fading is neglected, then it is straightforward to calculate the expected value of the random quantities in the coupling equations via the normalized user load intensities. The following lemma is a special case of a known statement on the moments of Poisson point processes (see, for example, *Klenke* (2006), Corollary 24.15):

Lemma 4.1: *If a random snapshot follows the restricted standard distribution on snapshots and the best server assumption holds, then the expected values of the coupling elements are*

$$\mathbb{E}(c_{ii}^{\uparrow}) = \int_{A_i} T_\ell^{\uparrow}(x)\mathrm{d}x, \qquad \mathbb{E}(c_{ij}^{\uparrow}) = \int_{A_j} \frac{\gamma_i^{\uparrow}(x)}{\gamma_j^{\uparrow}(x)} T_\ell^{\uparrow}(x)\mathrm{d}x$$

$$\mathbb{E}(c_{ii}^{\downarrow}) = \int_{A_i} \omega(x) T_\ell^{\downarrow}(x)\mathrm{d}x, \qquad \mathbb{E}(c_{ij}^{\downarrow}) = \int_{A_i} \frac{\gamma_j^{\downarrow}(x)}{\gamma_i^{\downarrow}(x)} T_\ell^{\downarrow}(x)\mathrm{d}x,$$

$$\mathbb{E}(p_i^{(\eta)}) = \int_{A_i} \frac{\eta^{\downarrow}(x)}{\gamma_i^{\downarrow}(x)} T_\ell^{\downarrow}(x)\mathrm{d}x.$$

for any network $\mathcal{N}$ and $i, j \in \mathcal{N}$, $i \neq j$. □

The lemma is intuitive and easy to see, if orthogonality loss, noise, and channel gain are step functions. This case is common in practice, because the data is usually specified over a discretization of the area into pixels (cf. p. 112). The lemma can be adapted to the case of cell association specified in terms of probability functions over the area. This (and the calculation of the cell-assignment probabilities) is the gist of the statistical estimation method introduced by *Türke* (2006). In the presence of shadowing, the accurate calculation of the expected coupling values (extended statistical method by *Türke*, 2006) is involved because the channel gain and the cell areas become dependent random variables.

4.2.2 Estimating performance indicators

The expected coupling matrix as calculated in Lemma 4.1 is used to compute estimates for all performance indicators defined in Sec. 3.4. The means of all cell-based performance indicators are estimated by using the expected coupling matrix elements in the definitions in Sec. 3.4.1.

We approximate the set of E_c/I_0-covered points by adapting the coverage threshold. For an accurate computation, the quantiles of the distribution of the highest E_c/I_0 are needed for each point. This distribution is in general different for different points in the planning area. Under the assumption that it is identical up to shifting the mean value, however, there is a global adjustment parameter ξ such that

$$\{x \in A \mid \mathbb{E}(\mathbf{1}(x)_{E_c/I_0}) \geq x\} = \{x \in A \mid \mathbb{E}(E_c/I_0) \geq \mu_{E_c/I_0} + \xi\},$$

which means that the quantiles can be determined by using the mean E_c/I_0 value. (E_c/I_0 is a shorthand for the right-hand side of (2.4.2) in the above formula.) Our strategy for approximating the set of covered points is to compute an estimate for the expected E_c/I_0 level based on expected coupling values and path loss attenuation, and to use an empirically calibrated threshold in the coverage condition.

Computational complexity. The expected coupling elements from Lemma 4.1 can be computed efficiently for the common representation of planning data. The planning area is usually discretize into rectangular pixels, and all functions appearing in the lemma are specified by explicitly stating the function value for each pixel. For computing the expected coupling elements, two steps are necessary: First, the cell areas are determined by identifying the cell with the highest channel gain for each pixel. Second, the referring integrals are computed as a sum over the pixels belonging to a certain cell area. This basically requires an effort proportional to the number of pixels times the number of cells. The input size of the path loss functions alone, however, is also proportional to the number of pixels times the number of cells. The computation time is thus basically linear in the input size.

The expected coupling matrix and the estimates based on it can be calculated efficiently also in a practical implementation. The last row in Tab. 4.1 shows that in the cases considered in our computational experiments, this took at most eleven seconds.

4.2.3 Error analysis in a simplified setting

We analyze the estimation error of expected interference coupling in a simplified setting based on the generalized pole equations introduced in Sec. 3.3. We consider a single reference service and a fixed estimate for the other-to-own-cell interference ratio. The interplay of partly correlated random variables then reduces to considering the number of users in a single cell, and both the precise mean of the performance indicators and their expected-coupling estimates can be calculated and compared explicitly. The experiments in Sec. 4.4.2 will show that the insights thus obtained apply in general.

Consider a single cell with a fixed average other-to-own-cell interference ratio $\hat{\iota}^{\downarrow}$ and assume that there is no noise. Suppose that there are m users, which all wish to access a reference service r with a user load factor of $\ell_r^{\downarrow}$. We assume that load control always admits the maximum number of users $m^* \leq m$ for which the system load is still within the desired bounds.

Calculating cell power and grade of service in the simplified setting. For m users in the system, the coupling element is calculated in the simplified setting as

$$c^{\downarrow}(m) = m^* \cdot \ell_r^{\downarrow} .$$

The load equation is

$$\bar{p}^{\downarrow}(m) = (1 + \hat{\iota}^{\downarrow}) c^{\downarrow}(m) \bar{p}^{\downarrow}(m) + p^{(c)} .$$

Serving the users therefore requires a power of

$$\bar{p}^{\downarrow}(m) = \frac{p^{(c)}}{1 - (1 + \hat{\iota}^{\downarrow}) \ell_r^{\downarrow} m^*} , \tag{4.3}$$

if this value is positive; otherwise there are too many users. The maximum feasible number of users is calculated by setting $\bar{p}^{\downarrow} = \bar{p}^{\downarrow}_{\text{max}}$ in (4.3) as

$$m^{(\text{max})} := \frac{\bar{p}^{\downarrow}_{\text{max}} - p^{(c)}}{\bar{p}^{\downarrow}_{\text{max}} (1 + \hat{\iota}^{\downarrow}) \ell_r^{\downarrow}} .$$

From the snapshot with m users, the system will thus admit

$$m^* = \min\big(\lfloor m^{(\text{max})} \rfloor, m\big)$$

users. The grade of service for m users is hence

$$\lambda^{\downarrow}(m) = \min(1, m^*/m) .$$

The load factor (see Sec. 3.3) before load control is

$$(1 + \hat{\iota}^{\downarrow}) \ell_r^{\downarrow} m .$$

Exact and estimated means of power and scaling. We assume the number of users to be Poisson distributed (see p. 29). We denote the parameter of the distribution, *i.e.*, the average number of users in the cell, by $\phi \geq 0$. As the number of users is the only random influence, the expected load factor is

$$\mathbb{E}\big((1+\hat{\imath}^{\downarrow})\ell_r^{\downarrow} m\big) = (1+\hat{\imath}^{\downarrow})\ell_r^{\downarrow}\phi\,, \tag{4.4}$$

and its variance is

$$\mathrm{Var}\big((1+\hat{\imath}^{\downarrow})\ell_r^{\downarrow} m\big) = (1+\hat{\imath}^{\downarrow})^2(\ell_r^{\downarrow})^2\phi\,. \tag{4.5}$$

We calculate the expected power and grade of service as

$$\mathbb{E}(\bar{p}^{\downarrow}) = \sum_{m \leq m^{(\max)}} P(m)\bar{p}^{\downarrow}(m) + \sum_{m > m^{(\max)}} P(m)\bar{p}^{\downarrow}(\lfloor m^{(\max)} \rfloor)\,, \tag{4.6a}$$

$$\mathbb{E}(\lambda^{\downarrow}) = \sum_{m \leq m^{(\max)}} P(m) \qquad + \sum_{m > m^{(\max)}} P(m)\lfloor m^{(\max)} \rfloor / m\,. \tag{4.6b}$$

The term $P(m)$ here denotes the probability that exactly m users are present. The latter components in both equations are infinite series that converge.

The mean of the coupling element, on the other hand, is $\mathbb{E}(c^{\downarrow}) = \phi\ell_r^{\downarrow}$. The expected-coupling estimates in the decoupled case are therefore calculated similar to Lemma 3.9 by distinguishing the cases of an average load factor above or below the critical value as

$$\bar{p}^{\downarrow} = \begin{cases} \frac{p^{(c)}}{1-(1+\hat{\imath}^{\downarrow})\ell_r^{\downarrow}\phi} & \text{if } \phi \leq m^{(\max)}, \\ \bar{p}^{\downarrow}_{\max} & \text{otherwise,} \end{cases} \tag{4.7a}$$

$$\lambda^{\downarrow} = \begin{cases} 1 & \text{if } \phi \leq m^{(\max)}, \\ \frac{\bar{p}^{\downarrow}_{\max} - p^{(c)}}{\bar{p}^{\downarrow}_{\max}(1+\hat{\imath}^{\downarrow})\ell_r^{\downarrow}\phi} & \text{otherwise.} \end{cases} \tag{4.7b}$$

Analysis results. The exact values (4.6) and the expected-coupling estimates (4.7) for mean cell power and grade of service are calculated for the parameters

$$p^{(c)} = 2\,\mathrm{W}, \qquad \bar{p}^{\downarrow}_{\max} = 14\,\mathrm{W}, \qquad \hat{\imath}^{\downarrow} = 1.08, \qquad \ell_r^{\downarrow} = 0.008, 0.059\,.$$

The first choice $\ell_r^{\downarrow} = 0.008$ corresponds to a voice service with 50 % activity; $\ell_r^{\downarrow} = 0.059$ is realistic for a video call service with 100 % activity. The parameters have furthermore been chosen such that $m^{(\max)}$ is integral to exclude rounding effects.

The results are depicted in Fig. 4.2. The abscissas in Fig. 4.2 show the mean load factor (before load control) as stated in (4.4). The curves differ only in the variance (4.5) of the load factor, which is determined by the service properties. The graphs are labeled by their *coefficient of variation* (c.o.v.), which is calculated for any random variable with non-zero mean as the ratio of its standard deviation over its

Figure 4.2: Average cell performance for different traffic variants

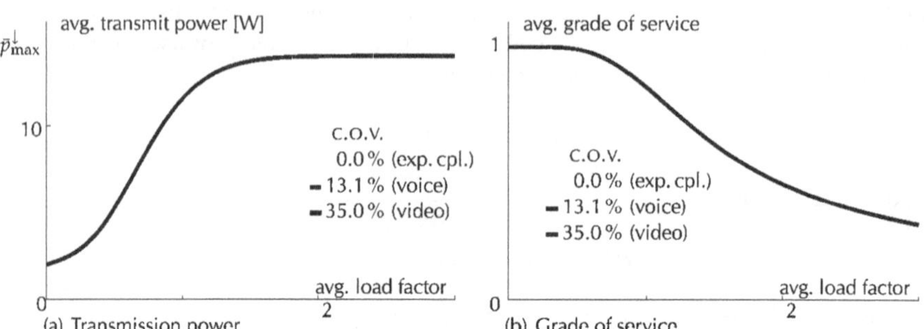

(a) Transmission power (b) Grade of service

mean (*Jain*, 1991). The light curve (c.o.v. of 0.0 %) represents the expected-coupling estimates. The voice and video services correspond to c.o.v. of 13.1 % and of 35.0 %.

The accuracy of the estimates for power are shown in Fig. 4.2(a). For a low average load factor, expected coupling underestimates the average transmit power because transmit power is a strictly convex function of the load factor below the critical value. For a low average load factor, the snapshots in which users are blocked have small probability, and the convex aspect prevails. The same effect has been noted as an underestimation bias by *Dziong et al.* (1999).

If the average load factor is high, on the other hand, the average power is overestimated by the expected-coupling approach. In the region with expected overload, the expected-coupling approximation disregards the fact that the snapshots for which the maximum power is not reached have a positive probability. This probability diminishes as the average load factors increases, so the precise means approach the estimate from below. This overestimation bias for high load factors has not been observed before because most previous investigations (*Dziong et al.*, 1999; *Staehle*, 2004) do not clip cell powers, but instead condition the analysis to the case that finite powers suffice. *Türke* (2006) clips the cell powers, so the effect is visible in his results, but it is not explicitly addressed.

The deviation of the precise values from the estimate is more pronounced the higher the variance of the load factor. If it had zero variance, the estimate would be exact; for speech users, small deviations occur; for video users inducing the highest variance of the load factor, the deviations are largest.

The analysis of the quality of power estimates applies basically also to the grade of service, but in an inverted fashion, as Fig. 4.2(b) shows. The estimate (4.7b) remains at one as long as the mean load factor is below the critical value. This estimate is quite accurate as long as the average load factor is low enough for blocking to occur with negligible probability. As the mean load factor approaches the critical value, the quality of the estimate deteriorates and overestimation is largest precisely at the critical load factor. Beyond the critical value, overestimation becomes underestimation, and the graphs for the precise means eventually approach the the estimate from above.

Figure 4.3: Average cell performance in the load range relevant for network planning

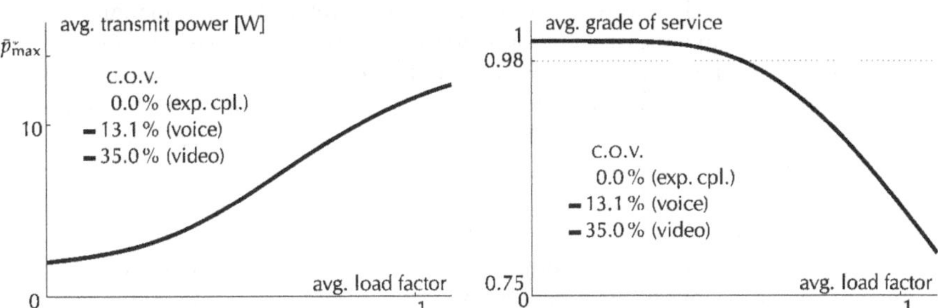

Usefulness of estimates in practice. Estimation quality behaves similar for power and grade of service, but the grade of service estimates are less useful in practice. Radio network fine planning usually deals not with pathological network configurations, but with reasonable ones. The task of optimization is to remedy capacity problems that are encountered in a small number of cells. In consequence, most cells have average load factors well below the critical value, and there are a few outliers.

The plots from Fig. 4.2 are repeated for the practice-relevant range of load factors in Fig. 4.3. The power estimates (Fig. 4.3(a)) become more accurate for a load factor approaching the critical value. The problems in overloaded cells are reliably detected by the conservative estimation bias. The estimates for grade of service (Fig. 4.3(b)), on the other hand, are worse for the most interesting cells, namely those close to overload.

The problem of inaccurate estimates for grade of service is severe because strict restrictions are typically imposed. Blocking rates of more than 2 % are usually considered unacceptable. As the plot shows, this threshold is exceeded by far at the critical load factor, even if only a speech service is considered. The estimate, however, does not indicate any problem. Precisely from which load factor on blocking becomes intolerable depends on the service mix.

4.3 Refined estimates for the expected grade of service

The expected-coupling approach has turned out to deliver estimates for the expected grade of service that are insufficient for most practical situations. A more refined technique is therefore needed; including second-order moments is the natural extension of the methods developed so far.

The traditional technique for including the variance of interference coupling is based on the pole equation. Again, using a fixed estimate for the other-to-own-cell interference ratio simplifies the computation of the grade of service considerably. Only stochastic variations of intra-cell interference are considered; inter-cell interference is accounted for as a multiple of intra-cell interference. *Viterbi & Viterbi* (1993) use essentially this method and demonstrate that it works well in a synthetic example scenario.

The classic approach, however, does not carry over well to practical planning cases, because there is no universally valid estimate for the other-to-own-cell interference ratio. In scenarios with irregular cell structure and inhomogeneous traffic distributions, there is a broad range of values, as Fig. 3.4 shows. A refined approach should hence rather use cell-specific estimates for the other-to-own-cell interference ratio.

In this section, we develop a refined estimation technique for the grade of service, which is based on a modification of this idea. For achieving better results, we untie the definition of other-to-own-cell interference ratio and calculate with individual coupling elements and transmit powers. To simplify the formulation, we use fixed estimates for the quantities governed by the traffic in the other cells. The quality of our refined estimates is investigated empirically and discussed in Sec. 4.4.2.

Refined estimates for grade of service in uplink. We start from a decomposition of the expected value of the scaling factor for a given cell, which is similar to Lemma 3.8. The following lemma basically states that for determining a cell's blocking ratio, we may calculate with the maximum power for this cell:

Lemma 4.2: *We are given a random coupling matrix $(c^{\uparrow}_{ij})_{i,j\in\mathcal{N}}$ with values in $\mathbb{R}^{\mathcal{N}\times\mathcal{N}}_{+}$. Let P be the joint probability distribution of the coupling elements $c^{\uparrow}_{ij}$, and suppose that the complementarity system (3.25) has a solution with probability one. Let furthermore $(\bar{p}^{\uparrow}_i)_{i\in\mathcal{N}}$ and $(\lambda^{\uparrow}_i)_{i\in\mathcal{N}}$ be defined as the induced random average cell powers and scaling values in a (uniquely determined) solution to (3.25). Let B be the set of outcomes for which $c^{\uparrow}_{ii} > 0$ and*

$$\bar{p}^{\uparrow}_{\max} c^{\uparrow}_{ii} \geq \bar{p}^{\uparrow}_{\max} - \eta^{\uparrow}_i - \sum_{j\neq i} \lambda^{\uparrow}_j c^{\uparrow}_{ij} \bar{p}^{\uparrow}_j \tag{4.8}$$

and let $\overline{B}$ be its complement. Then

$$\mathbb{E}(\lambda^{\uparrow}_i) = \int_{\overline{B}} 1 \mathrm{d}P + \int_B \frac{\bar{p}^{\uparrow}_{\max} - \eta^{\uparrow}_i - \sum_{j\neq i} \lambda^{\uparrow}_j c^{\uparrow}_{ij} \bar{p}^{\uparrow}_j}{\bar{p}^{\uparrow}_{\max} c^{\uparrow}_{ii}} \mathrm{d}P\,. \tag{4.9}$$

Proof. We show that the integrands in (4.9) correspond to the scaling value $\lambda^{\uparrow}_i$ under the respective conditions. First, we see that $\lambda^{\uparrow}_i = 1$ whenever (4.8) is violated. In fact, if $\lambda^{\uparrow}_i < 1$, then

$$\bar{p}^{\uparrow}_i \overset{(3.23)}{=} \bar{p}^{\uparrow}_{\max} \overset{(3.21)}{=} \lambda^{\uparrow}_i c^{\uparrow}_{ii} \bar{p}^{\uparrow}_{\max} + \textstyle\sum_{j\neq i} \lambda^{\uparrow}_j c^{\uparrow}_{ij} \bar{p}^{\uparrow}_j + \eta^{\uparrow}_i \overset{\lambda^{\uparrow}_i<1}{<} c^{\uparrow}_{ij} \bar{p}^{\uparrow}_{\max} + \textstyle\sum_{j\neq i} \lambda^{\uparrow}_j c^{\uparrow}_{ij} \bar{p}^{\uparrow}_j + \eta^{\uparrow}_i\,,$$

so (4.8) cannot be violated.

Now suppose that (4.8) holds. Then, we check that $\bar{p}^{\uparrow}_i = \bar{p}^{\uparrow}_{\max}$: If $\lambda^{\uparrow}_i < 1$, then $\bar{p}^{\uparrow}_i = \bar{p}^{\uparrow}_{\max}$ by complementarity, *i. e.*, implication (3.23). If $\lambda^{\uparrow}_i = 1$, this means that in addition to

$$\bar{p}^{\uparrow}_{\max} c^{\uparrow}_{ii} \geq \bar{p}^{\uparrow}_{\max} - \eta^{\uparrow}_i - \textstyle\sum_{j\neq i} \lambda^{\uparrow}_j c^{\uparrow}_{ij} \bar{p}^{\uparrow}_j$$

we have, by (3.22),

$$c_{ii}^{\uparrow}\bar{p}_i^{\uparrow} = \bar{p}_i^{\uparrow} - \sum_{j\neq i}\lambda_j^{\uparrow}c_{ij}^{\uparrow}\bar{p}_j^{\uparrow} - \eta_i^{\uparrow}\,.$$

By subtracting the equation from the inequality, we obtain

$$c_{ii}^{\uparrow}(\bar{p}_{\max}^{\uparrow} - \bar{p}_i^{\uparrow}) \geq (\bar{p}_{\max}^{\uparrow} - \bar{p}_i^{\uparrow})\,.$$

Since we assumed $\lambda_i^{\uparrow} = 1$, however, a nonnegative solution to the coupling system can only exist if $c_{ii}^{\uparrow} < 1$, so $\bar{p}_i^{\uparrow} = \bar{p}_{\max}^{\uparrow}$ must hold also here.

By substituting this into (3.21) and rearranging, we obtain

$$\lambda_i^{\uparrow} = \frac{\bar{p}_{\max}^{\uparrow} - \eta_i^{\uparrow} - \sum_{j\neq i}\lambda_j^{\uparrow}c_{ij}^{\uparrow}\bar{p}_j^{\uparrow}}{\bar{p}_{\max}^{\uparrow}c_{ii}^{\uparrow}}\,,$$

which means that the second integrand corresponds to $\lambda_i^{\uparrow}$ over the set B. □

For a reasonably planned network in practice, the condition that a solution exists almost always is not restrictive. A unique solution can, for example, be obtained as the lexicographically minimal one. In practical experiments, we used the outcome of the heuristic (3.31), which always converged during simulations.

The form (4.9) is still complex, but it facilitates an estimate of the expected grade of service that can be efficiently calculated. Our strategy is to approximate the set described by (4.8) and the second integrand in (4.9) with expressions that only depend on $c_{ii}^{\uparrow}$, *i.e.*, the users in the own cell. To this end, we plug in expected-coupling estimates $\sum_{j\neq i}\tilde{\lambda}_j^{\uparrow}\tilde{c}_{ij}^{\uparrow}\tilde{p}_j^{\uparrow}$ for terms related to the behavior of other cells.

Under the restricted standard random model specified in Def. 4 on p. 30, the only remaining random variable $\bar{p}_{\max}^{\uparrow}c_{ii}^{\uparrow}$ is a weighted sum of as many Poisson variables as there are services. Because this distribution has no simple description, we approximate it with a Gaussian variable with mean and variance

$$\mathbb{E}(\bar{p}_{\max}^{\uparrow}c_{ii}^{\uparrow}) = \bar{p}_{\max}^{\uparrow}\mathbb{E}(c_{ii}^{\uparrow})\,,$$
$$\mathrm{Var}(\bar{p}_{\max}^{\uparrow}c_{ii}^{\uparrow}) = (\bar{p}_{\max}^{\uparrow})^2\sum_r(\ell_r^{\uparrow})^2\int_{A_i}T_r(x)\mathrm{d}x\,.$$

The normal approximation is better if the average number of users in the cell is large. The expected coupling element $\mathbb{E}(c_{ii}^{\uparrow})$ is calculated using Lemma 4.1; the identity for variance, again, follows by approximating the involved functions with step functions.

For the estimated blocking limit in uplink defined as

$$\kappa_i^{\uparrow} := \bar{p}_{\max}^{\uparrow} - \eta_i^{\uparrow} - \sum_{j\neq i}\tilde{\lambda}_j^{\uparrow}\tilde{c}_{ij}^{\uparrow}\tilde{p}_j^{\uparrow}\,,$$

we consider the probability density function ψ_i of $\bar{p}_{\max}^{\uparrow}c_{ii}^{\uparrow}$ and estimate $\mathbb{E}(\lambda_i^{\uparrow})$ as

$$\mathbb{E}(\lambda_i^{\uparrow}) \simeq \int_0^{\kappa_i^{\uparrow}}\psi_i(x)\mathrm{d}x + \int_{\kappa_i^{\uparrow}}^{\infty}\frac{\kappa_i^{\uparrow}}{x}\psi_i(x)\mathrm{d}x\,, \tag{4.10}$$

provided that $\kappa_i^{\uparrow} > 0$. Otherwise, we estimate the grade of service as zero. This case, however, would indicate poor network design and did not occur in the computational tests presented in Sec. 4.4.2 below. The integral (4.10) is computed numerically.

Refined estimates for the grade of service in downlink. A similar approach works for downlink. Because the coupling matrix is now constructed row wise, however, we take into account the stochastics not only of the main diagonal element but of an entire matrix row.

Lemma 4.3: *We are given random coupling elements $(c_{ij}^{\downarrow})_{i,j}$ that form a random matrix with values in $\mathbb{R}_{+}^{\mathcal{N}\times\mathcal{N}}$. Let their joint probability distribution be P. Let furthermore the random variables $(\bar{p}^{\downarrow})_{i\in\mathcal{N}}$ and $(\lambda_i^{\downarrow})_{i\in\mathcal{N}}$ be defined as the average cell powers and scaling values in the solution to the complementarity system (3.26); they are functions of the random matrix. Let B be the set of outcomes for which*

$$c_{ii}^{\downarrow}\bar{p}_{\max}^{\downarrow} + \sum_{j\neq i} c_{ij}^{\downarrow}\bar{p}_j^{\downarrow} + p_i^{(\eta)} \geq \bar{p}_{\max}^{\downarrow} - p_i^{(c)} \tag{4.11}$$

and let $\overline{B}$ be its complement. Then for any $i \in \mathcal{N}$

$$\mathbb{E}(\lambda_i^{\downarrow}) = \int_{\overline{B}} 1 \mathrm{d}P + \int_B \frac{\bar{p}_{\max}^{\downarrow} - p_i^{(c)}}{c_{ii}^{\downarrow}\,\bar{p}_{\max}^{\downarrow} + \sum_{j\neq i} c_{ij}^{\downarrow}\bar{p}_j^{\downarrow} + p_i^{(\eta)}} \mathrm{d}P .$$

Proof. The argument for verifying the downlink decomposition is similar to the proof of Lemma 4.2 for the uplink. If $\lambda_i^{\downarrow} < 1$, we have

$$\bar{p}_i^{\downarrow} = \bar{p}_{\max}^{\downarrow} = \lambda_i^{\downarrow}(c_{ii}^{\downarrow}\,\bar{p}_{\max}^{\downarrow} + \sum_{j\neq i} c_{ij}^{\downarrow}\bar{p}_j^{\downarrow} + p_i^{(\eta)}) + p_i^{(c)} < c_{ii}^{\downarrow}\,\bar{p}_{\max}^{\downarrow} + \sum_{j\neq i} c_{ij}^{\downarrow}\bar{p}_j^{\downarrow} + p_i^{(\eta)} + p_i^{(c)} ,$$

so (4.11) is not violated. If, on the other hand, (4.11) holds, then similarly to the argument for the uplink, we can see that $\bar{p}_i^{\downarrow} = \bar{p}_{\max}^{\downarrow}$ and substitute this into (3.22) to see that the integral is correctly computed. □

In the formulas in the lemma, the power values $\bar{p}_j^{\downarrow}$ are the only influences stemming from cells other than i. We will estimate these by fixed approximate values $\tilde{p}_j^{\downarrow}$. As in the uplink, we approximately calculate the means by numerically integrating a Gaussian random variable with parameters

$$\mathbb{E}\Big(c_{ii}^{\downarrow}\tilde{p}_{\max}^{\downarrow} + \sum_{j\neq i} c_{ij}^{\downarrow}\tilde{p}_j^{\downarrow} + p_i^{(\eta)}\Big) = \mathbb{E}(c_{ii}^{\downarrow})\tilde{p}_{\max}^{\downarrow} + \sum_{j\neq i}\mathbb{E}(c_{ij}^{\downarrow})\tilde{p}_j^{\downarrow} + \mathbb{E}(p_i^{(\eta)}) ,$$

$$\mathrm{Var}\Big(c_{ii}^{\downarrow}\tilde{p}_{\max}^{\downarrow} + \sum_{j\neq i} c_{ij}^{\downarrow}\tilde{p}_j^{\downarrow} + p_i^{(\eta)}\Big) = \sum_r \int_{A_i}\Big(\omega(x) + \frac{\gamma_j^{\downarrow}(x)}{\gamma_i^{\downarrow}(x)} + \frac{\eta^{\downarrow}(x)}{\gamma_i^{\downarrow}(x)}\Big)^2 \ell_r^{\downarrow}(x)^2 T_r(x)\mathrm{d}x .$$

We denote its probability density function by ψ_i and calculate estimates as

$$\mathbb{E}(\lambda_i^{\downarrow}) \simeq \int_0^{\bar{p}_{\max}^{\downarrow} - p_i^{(c)}} \psi_i(x)\mathrm{d}x + \int_{\bar{p}_{\max}^{\downarrow} - p_i^{(c)}}^{\infty} \frac{\bar{p}_{\max}^{\downarrow} - p_i^{(c)}}{x}\psi_i(x)\mathrm{d}x .$$

The integral is, again, calculated using numerical integration.

In general, we are in a less fortunate position in the downlink than in the uplink, because the influence of surrounding cells is inherently higher due to orthogonality.

This makes estimates relying on the state of the current cell only less precise, but there is an additional trick for improving precision by a smarter choice of the values $\bar{p}_j^{\downarrow}$. The first candidates for the estimates are their means, *i. e.*, $\bar{p}_j^{\downarrow} = \mathbb{E}(\bar{p}_j^{\downarrow})$. However, we consider the case in which cell i blocks. It is thus intuitively clear that we obtain better results if we use the means $\mathbb{E}(\bar{p}_j^{\downarrow} | \bar{p}_i^{\downarrow} = \bar{p}_{\max}^{\downarrow})$, *i. e.*, restricted to the blocking case. We cannot easily calculate these means, of course, but there is a simple tweak to the expected-coupling method for producing an estimate: we fix $\bar{p}_i^{\downarrow} = \bar{p}_{\max}^{\downarrow}$ and solve a complementarity system (3.26) of reduced dimension for the residual power variables. This entails an additional effort of solving n complementarity systems of size $n-1$, but it improves the quality of the estimates considerably (*Ryll*, 2006).

4.4 Computational experiments

We now investigate the validity of our models and estimation schemes on realistic data instances. First, the assumption of perfect load control is scrutinized in Sec. 4.4.1. We check for a large number of random snapshots how close the results of discrete load control decisions come to the idealizing assumptions. Second, we compare the methods developed in this chapter to simplified Monte Carlo simulation in Sec. 4.4.2 to determine how accurately they estimate mean values of the performance indicators. We treat only the downlink results here, as it is the bottleneck direction in most traffic scenarios and the direction for which we conduct optimization in the next chapter. We use the planning scenarios for Berlin and Lisbon from the publicly available data provided by the MOMENTUM *Project* (2003). The data is discussed in detail in Sec. 5.4.1.

We vary the traffic intensity, the service mix, and the network configuration to obtain results for settings with various characteristics. There are three traffic levels: low, medium, and high intensity. With low intensity, hardly any cells are ever in overload; with medium intensity, some cells are typically in overload; with high traffic intensity, which is only used in the second part, overload is a frequently occurring event. Besides the full service mixes comprising speech telephony, video telephony, and various data services, we consider two restricted variants containing only speech and only video users. The traffic intensity is scaled in these cases such that the normalized expected user load remains the same; the factors are listed in Tab. A.1 in the appendix. Details on the parameters for the different services are given in Tab. A.6 in the appendix. We use a non-optimized network design with regular parameter settings for Berlin and Lisbon, and three different optimized configurations for Berlin and one for Lisbon. In general, the optimized configurations use fewer sites and cells than the regular ones and are tuned for low interference coupling.

For both sets of experiments, we first carry out an in-depth analysis of results on the example network "Berlin optimized 1" and then present aggregated results for the residual configurations to demonstrate that the findings are representative. The computations have been made with the custom software described in Ch. 1 and depicted in Fig. 1.2.

4.4.1 Discrete load control

To show that perfect load control is a reasonable assumption, we strive to identify a classical discrete load control scheme that leads to a network performance close to the performance indicators computed by our compact system model. A discrete load control scheme decides for a given snapshot which users are granted access to the network and which users are blocked. For each evaluated cell, we compare two pairs of numbers: First, the transmit powers needed to serve the admitted users are compared to the power variables computed by the complementarity model. Second, the ratio of served traffic load over total traffic load in the snapshot is compared to the scaling factors. If the deviations in both cases remain tolerable for a given discrete scheme, this justifies disregarding individual users. For power values, any load-control scheme should ideally adjust the transmit powers of a load-controlled cell to the value $\bar{p}^{\downarrow}_{\max}$. If it is exceeded, the cell is still in overload; otherwise, capacity is wasted. In this light, we will also refer to $\bar{p}^{\downarrow}_{\max}$ as the *target value* below.

We evaluate 750 snapshots separately in each case. For each snapshot, we compute the solutions to the system model (3.25) and (3.26). Subsequently, we use the resulting scaling values in the methods proposed in Sec. 3.2.4 to determine subsets of all users in the snapshot. We then compute the solution to the classical, unscaled interference coupling systems (3.5) and (3.14) for these subsets.

We test the four load-control schemes described in Sec. 3.2.4: *Random Activation* serves each mobile with a probability of the scaling factor applicable to the requested cell. The *Random Order* scheme removes mobiles until the sum of row elements is reduced below the row sum of the scaled matrix. The *Knapsack* scheme follows the same approach as Random Order but tries to meet the bound on the row sum with least possible slack. The *Multiple Knapsack* scheme reduces all matrix elements separately to at most the levels from the scaled solution, which guarantees that the maximum cell-load is never exceeded. All schemes reject users only in load-controlled cells (*i. e.*, cells with a scaling values smaller than one).

Comparison of different discrete load control schemes. The effect of the different approaches to load control are illustrated for the example network in the histograms in Fig. 4.4. All figures show the histogram of cell loads over all snapshots for the complete service mix at medium traffic intensity. The shaded areas correspond to load-controlled cells. The maximum average power $\bar{p}^{\downarrow}_{\max} = 14\,\mathrm{W}$ is indicated by the light dotted line.

Figs. 4.4(a) and 4.4(b) contrast the absence of any load control mechanism with perfect load control: If the user load is not reduced (Fig. 4.4(a)), many cells exceed the average power limit $\bar{p}^{\downarrow}_{\max}$. In about 10 % of all cases, even the nominal maximum power $p^{\downarrow}_{\max} = 20\,\mathrm{W}$ is attained. (Transmit power was clipped at the maximum nominal value in all cases, thence the concentration on the last bar.) With perfect load control (Fig. 4.4(b)), all load-controlled cells are exactly at the maximum average power $\bar{p}^{\downarrow}_{\max}$.

For the discrete load control schemes, the transmit powers of load-controlled cells

Figure 4.4: Comparison of discrete load-control schemes—distribution of average cell transmit powers (Network "Berlin optimized 1", medium traffic intensity, complete service mix, downlink).

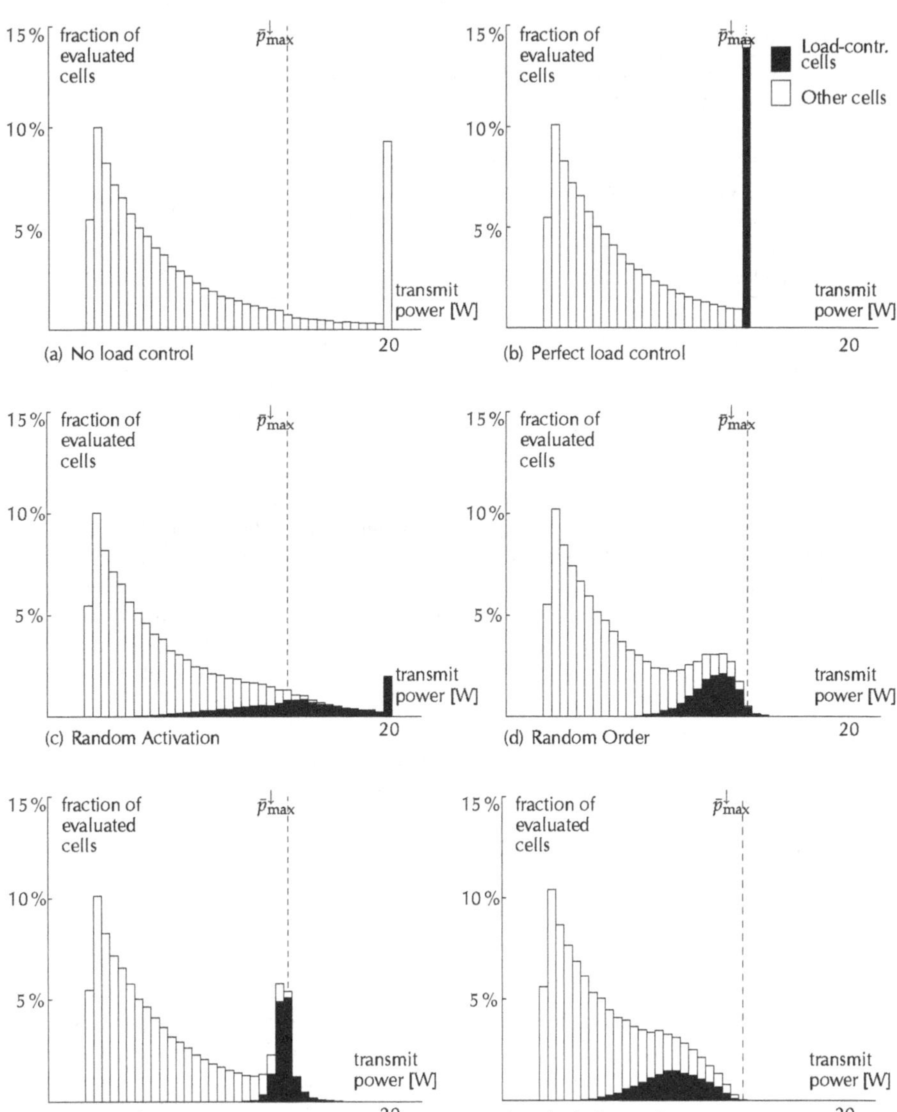

(a) No load control

(b) Perfect load control

(c) Random Activation

(d) Random Order

(e) Knapsack

(f) Multiple Knapsack

Table 4.2: Perfect vs. discrete load-control: transmit power deviations

			speech service			all services		
intensity	network	# eval[a]	% lc[b]	$\lvert\Delta\bar{p}\rvert$ lc[c]	$\lvert\Delta\bar{p}\rvert$ tot[d]	% lc[b]	$\lvert\Delta\bar{p}\rvert$ lc[c]	$\lvert\Delta\bar{p}\rvert$ tot[d]
low	Lisbon reg.	123 000	0.01	0.19	0.000	0.55	0.59	0.005
	Lisbon opt.	96 000	0.00	0.00	0.000	0.30	0.58	0.002
	Berlin reg.	144 750	0.47	0.34	0.003	1.69	0.61	0.016
	Berlin opt.	91 500	0.03	0.11	0.000	2.48	0.52	0.019
medium	Lisbon reg.	123 000	3.50	0.29	0.016	4.52	0.64	0.041
	Lisbon opt.	96 000	2.68	0.24	0.010	5.32	0.55	0.041
	Berlin reg.	144 750	5.86	0.41	0.037	6.93	0.58	0.056
	Berlin opt.	91 500	10.64	0.17	0.027	13.89	0.47	0.089

[a] Number of evaluated cases (750 times the number of cells in the network)
[b] Fraction of load-controlled cases over all evaluated cases
[c] Mean absolute deviation from target power $\bar{p}^{\downarrow}_{\text{max}}$ in load-controlled cells [W]
[d] Mean absolute difference between power values for perfect and discrete load control in all cells [W]

are distributed around the target value $\bar{p}^{\downarrow}_{\text{max}}$ with varying characteristics. The "Random Activation" scheme is shown in Fig. 4.4(c). In general, the dispersion of cell loads around the desired value is roughly symmetric and large; also here many cells hit the absolute, nominal maximum power of 20 W. The scheme apparently does not reduce the load sufficiently. The "Random Order" scheme depicted in Fig. 4.4(d) seems to work better: There are hardly any occurrences of cells exceeding the target power value. However, there is a considerable tail of load controlled cells with transmit power values in the medium range. This is a weakness, because resources are wasted. The fault is largely eliminated by the "Knapsack" scheme (Fig. 4.4(e)). The power of blocking cells is more focused around the target value here. A few cells are still mildly in overload. The "Multiple Knapsack" scheme depicted in Fig. 4.4(f) is more conservative and reliably drops enough users, but at the price of more wasted capacity.

Among the four candidates, the "Knapsack" scheme best strikes the balance between reducing the cell load to the feasible region and packing cells for maximum resource usage. Detailed numbers substantiating this conclusion and the impressions from the figures are contained in Tabs. B.4 and B.5 in the appendix. We will therefore only report on the performance of this scheme on the remaining network instances.

Results for the Knapsack scheme for different networks and scenarios. The deviations of the discrete "Knapsack" load-control scheme from perfect load control in all four considered settings and for two service mixes are specified in Tab. 4.2 for transmit powers and Tab. 4.3 for grade of service. In both tables, we first list the number of evaluated cases for each network in column "# eval". For each combination of traffic intensity and service mix, we then specify the fraction of load-controlled cells (column "% lc"). For the power values, Tab. 4.2 states the average absolute deviation of the

Table 4.3: Perfect vs. discrete load control: differences in grade of service

			speech service		all services	
intensity	network	# eval[a]	% lc[b]	$\lvert\Delta\lambda\rvert$[c]	% lc[b]	$\lvert\Delta\lambda\rvert$[c]
low	Lisbon reg.	123 000	0.01	1.1	0.55	2.4
	Lisbon opt.	96 000	0.00	0.0	0.30	2.1
	Berlin reg.	144 750	0.47	2.5	1.69	4.1
	Berlin opt.	91 500	0.03	1.2	2.48	3.5
medium	Lisbon reg.	123 000	3.50	2.3	4.52	3.4
	Lisbon opt.	96 000	2.68	1.4	5.32	2.7
	Berlin reg.	144 750	5.86	3.8	6.93	4.5
	Berlin opt.	91 500	10.64	2.7	13.89	4.0

[a] Number of evaluated cases (750 times the number of cells in the network)
[b] Fraction of load-controlled cases over all evaluated cases
[c] Mean absolute difference between served traffic for perfect and discrete load control, load-controlled cells only [percentage points]

transmit power in load-controlled cells from the target value $\bar{p}^{\downarrow}_{\text{max}}$ in column "$|\Delta\bar{p}|$ lc"; column "$|\Delta\bar{p}|$ tot" lists the average absolute difference between transmit powers under perfect and discrete load control. For the grade of service values, Tab. 4.3 specifies the average absolute difference between the scaling factor and the fraction of normalized traffic served after discrete load-control in column "$|\Delta\lambda|$".

The results show that the difference between perfect load control and the discrete scheme are tolerable in all cases; furthermore, the deviation depends on the traffic intensity and on the service mix. Load-controlled cells are on average at most 0.41 W away from the target value of $\bar{p}^{\downarrow}_{\text{max}}$ if only speech users are present. For the complete service mix, this increases to at most 0.64 W. This difference is explained by the coarser average granularity of users in the latter case, which makes it more difficult to fill a cell up to its capacity. The overall difference in power values (including cells that are not load-controlled) is negligible in all cases, as its expected absolute value is at most 0.089 W and this value is only attained if blocking occurs frequently. For scaling values, the picture is similar. For fine-granular speech users, the expected estimation error of perfect load control is at most 2.7 percentage points; for the complete service mix, the expected difference is at most 4.5 percentage points. Given the fact that a typical single data user occupies about 5–10 % of the total cell capacity (cf. Tab. 2.1), much smaller values could not have been hoped for.

4.4.2 Precision of expected-coupling estimates

We compare the newly developed estimates for performance measures linked to capacity and transmit powers to the results of a simplified Monte Carlo analysis. The average transmit power and the grade of service are directly involved in the interference-coupling complementarity system and are therefore investigated in most detail; the

Figure 4.5: Transmit power estimates, medium traffic intensity

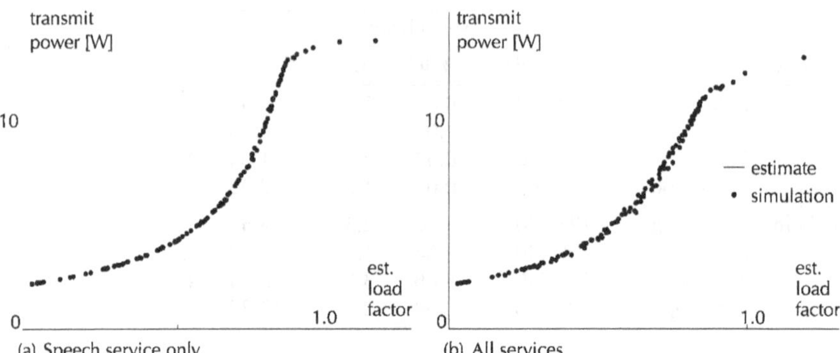

(a) Speech service only

(b) All services

estimation quality for average cell load, total transmit power, and the total grade of service can be directly deduced. Subsequently, we briefly review the quality of estimates for the remaining performance indicators.

Accurate expected values of the grade of service and transmit power are calculated as described in Sec. 4.1; they are accurate by at least 1 % with 99 % probability. We will refer to the values derived from the expected interference-coupling model as the *estimated* values and to the simulation results as the *observed* values. We first restrict the underlying random model to exclude shadow fading; the prerequisites of Lemma 4.1 are thus fulfilled and we use the precise mean of the coupling matrix. The impact of shadow fading is subsequently investigated separately. Recall the notions of the load factor and critical load factors introduced in Sec. 3.3. In the following, the *load factor* refers to the estimated expected load factor unless otherwise noted.

Transmit power estimates. The plots in Fig. 4.5 depict the quality of expected-coupling estimates for transmit powers at medium traffic intensity for the Berlin network; Fig. 4.5(a) contains the results for speech service only and Fig. 4.5(b) includes the complete service mix. The abscissas denotes the estimate of the downlink load factor $(1+\bar{t}_i^{\downarrow})\bar{c}_{ii}^{\downarrow}$ derived from the complementarity system using the expected coupling matrix; our estimate for the average output power of each cell as specified in Lemma 3.9 is reproduced in the plot by the light solid line. Each dot in the plots corresponds to one of the 122 cells; the ordinate represents the mean observed transmit power. The estimation error thus corresponds to the deviation of the dots from the solid line.

Both the pattern of the simulation results within each plot and the differences between the two plots are consistent with the results of the simplified analysis conducted in Sec. 4.2.3 above. In the MOMENTUM scenarios, the coupling value in all cells have a similar variance, so the dots lie almost on a line comparable to the ones in Fig. 4.2(a). For small load factors, the simulation results slightly exceed the estimate. As the load factor increases, the expected-coupling method tends to underestimate

Table 4.4: Accuracy of estimates for network "Berlin optimized 1"

		transmit power $\bar{p}^{\downarrow}$			grade of service $\lambda^{\downarrow}$ [a]		
intensity		speech	video	all	speech	video	all
low	max underestimation[b]	0.518	0.830	0.905	0.0	0.0	0.0
	max overestimation[b]	0.000	0.000	0.000	0.1	1.0	1.2
	max rel. error [%]	5.5	11.5	12.7	0.1	1.0	1.3
	mean rel. error [%]	2.0	5.8	6.7	0.0	0.1	0.1
	1−correlation	0.000	0.002	0.002	0.002	0.014	0.016
medium	max underestimation[b]	0.365	0.612	0.645	0.6	4.6	4.9
	max overestimation[b]	1.025	2.597	2.808	0.7	0.4	0.5
	max rel. error [%]	8.0	22.8	25.2	0.8	5.9	6.3
	mean rel. error [%]	1.8	6.2	7.2	0.1	0.4	0.5
	1−correlation	0.002	0.012	0.024	0.001	0.002	0.003
high	max underestimation[b]	0.328	0.711	0.770	0.8	3.2	3.9
	max overestimation[b]	0.966	2.614	2.948	0.6	0.3	0.4
	max rel. error [%]	7.4	23.4	26.8	1.2	4.5	5.4
	mean rel. error [%]	1.3	5.9	6.9	0.1	0.9	1.0
	1−correlation	0.001	0.008	0.010	0.000	0.001	0.001

[a] Refined estimates (Sec. 4.3)
[b] Unit: [W] for transmit power, [percentage points] for grade of service

the average power. The estimation error is largest at the critical load factor. In the case of the full service mix, the estimation error is generally larger, and the deviation is more pronounced. The reason is the coarser granularity of users, which increases the variance of interference coupling.

The computational results for all combinations of service mixes and traffic intensities are listed in the left part of Tab. 4.4. In each case, we list five key values describing the relation between the vectors of estimated and observed powers. First, the maximum absolute deviation of the observed value from the estimate per cell is specified; errors due to overestimation and underestimation are separated. Subsequently, the maximum relative error and the average relative error over all cells is shown. Finally, we indicate the complement of the correlation between the vectors of estimated and observed power values; the closer this value is to zero, the better is the relation between the two vectors described as a linear one. Small numbers are thus better throughout the table.

The numbers in Tab. 4.4 extend the observations from the discussion of Fig. 4.5. The comparison between the different service mixes is easily grasped: The variance of the elements of the coupling matrix increases from speech over video to the complete service mix. In consequence, the numbers increase in each row, *i. e.*, the estimates are worse for service mixes including more demanding individual users. The relation of prediction accuracy and traffic level is more subtle, but analogous to the analysis

Figure 4.6: Estimating the grade of service (complete service mix at medium traffic intensity)

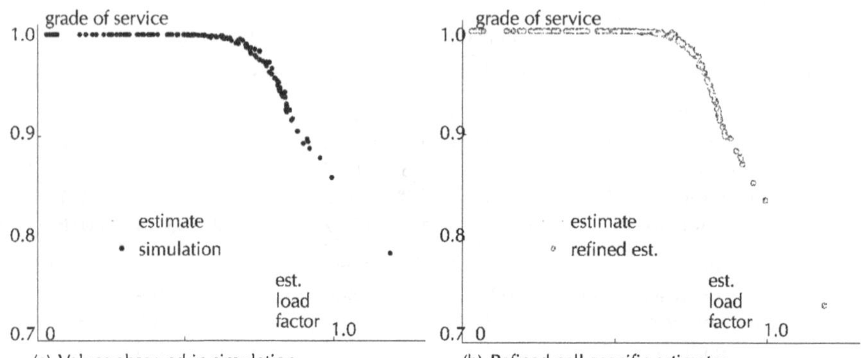

(a) Values observed in simulation (b) Refined cell-specific estimates

in Sec. 4.2.3: For low traffic intensity, we operate at the lower end of the graph, so the power levels are underestimated and overestimation is zero throughout. For higher traffic levels, the maximum underestimation is smaller, because now also overestimation occurs and both extremes cancel each other out to a certain degree. The largest differences between predicted and observed values are consistently overestimates in all cases except for that of low traffic intensity, where overestimation does not occur at all. This and the fact that the mean relative error and correlation values are worse for the medium traffic intensity (rather than for the high one) confirms our above observation: the accuracy of expected-coupling estimates is worst around the critical load factor. The medium traffic intensity has most cells operating in the vicinity of the critical load factor, so estimates deviate more than for high and low traffic intensities.

Estimates for grade of service. Observed values and estimates for grade of service for the complete service mix at medium traffic intensity are depicted in Fig. 4.6. Fig. 4.6(a) shows the observed values together with the expected coupling estimates. The pattern behaves as expected: the estimate remains at 1.0 up to the critical load factor. In simulation, however, blocking is observed long before reaching the critical load factor on average. The largest estimation error occurs at the critical load factor. The plot reveals the fundamental problem of expected-coupling estimates: The increasing blocking ratios of cells operating close to but below the critical load factor are not detected. Tab. B.3 in the appendix gives a detailed analysis of the accuracy of expected-coupling for the grade of service. The bottom line is that expected-coupling estimates convey relevant information only for low and high, but not for medium traffic intensities. The information that blocking virtually does not occur for low traffic intensities, however, is obvious also without any estimation technique.

The refined estimates indicate blocking already before the critical load factor, as Fig. 4.6(b) illustrates. The two plots in Figs. 4.6(a) and (b) are not identical, but the estimates seem to convey relevant information on which cells reject users. That the

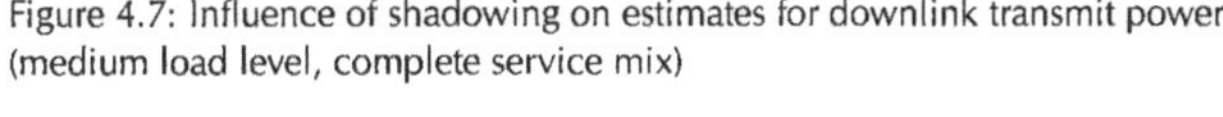

Figure 4.7: Influence of shadowing on estimates for downlink transmit power (medium load level, complete service mix)

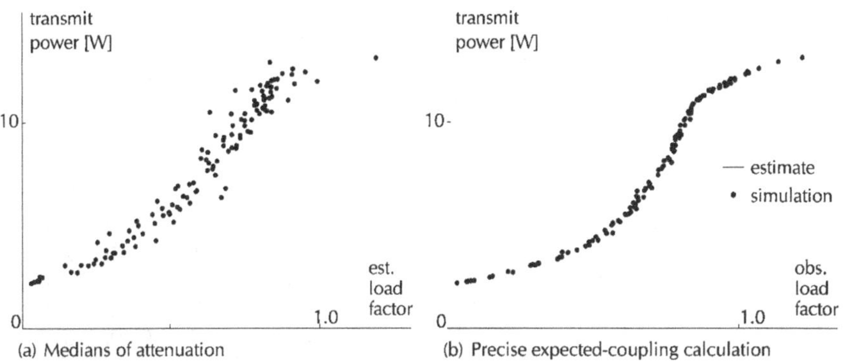

(a) Medians of attenuation

(b) Precise expected-coupling calculation

refined estimates improve estimation accuracy substantially is confirmed in detail by the results listed in Tab. B.3. The numerical analysis of the accuracy of refined grade of service estimates is contained in the right part of Tab. 4.4. Overestimation and underestimation follow a similar pattern as the power estimates, but they have exchanged roles: There is no underestimation for low traffic intensities; for medium and high traffic intensities, however, underestimation is the largest source of error. Overestimation is reduced to below 0.7 percent points in all cases and underestimation prevails; the refined estimation approach therefore has a pessimistic bias.

The effect of shadowing. We now extend the random model for generating snapshots to include shadow fading, and we are interested in how well the simple estimates based on medians of attenuation approximate the results of simulation in this case. Fig. 4.7(a) depicts the results for power estimates under medium traffic intensity and considering all services. The load factors are estimated as before, but the observed transmit power is now determined using simulation including shadowing. Geometrically, the abscissas of the 122 points in Fig. 4.7(a) are identical to the ones in Fig. 4.5(b), only the ordinates have changed; rearranging the values thus disperses the results to a diffuse cloud rather than a common curve. Although the general tendency of estimates indicated by the solid line is valid, the quality of individual estimates varies.

The figures in the left part of Tab. 4.5 substantiate the impression from the graphs. On the left-hand side, we see the analysis of our medians-of-attenuation estimates against the simulation results including shadowing. The estimates deviate more from the observed values than for simulation without shadowing. The overestimation and underestimation pattern are qualitatively different from the no-shadowing case: At low traffic intensity, there is now some overestimation. For medium and high traffic intensities, underestimation now prevails over overestimation, so the estimation has changed its bias. Moreover, the influence of the service mix is not as clear as before. Estimates for the speech-only scenario are about as good as for the complete service

Table 4.5: Quality of transmit power estimates in the presence of shadow fading

		medians-of-att. estimate			exp.-coupl. estimate		
intensity		speech	video	all	speech	video	all
low	max underestimation [w]	4.042	3.666	3.560	0.333	0.724	0.738
	max overestimation [w]	0.280	0.106	0.048	0.798	0.202	0.472
	max rel. error [%]	40.8	41.0	41.3	8.9	10.2	10.2
	mean rel. error [%]	15.5	17.9	18.5	1.4	5.8	6.0
	1−correlation	0.050	0.046	0.047	0.002	0.003	0.005
medium	max underestimation [w]	6.223	4.839	4.679	0.112	0.520	0.553
	max overestimation [w]	1.109	2.653	2.900	1.184	2.659	2.914
	max rel. error [%]	52.8	45.5	44.6	9.3	23.7	26.4
	mean rel. error [%]	19.1	17.4	17.4	1.8	7.7	9.1
	1−correlation	0.061	0.066	0.067	0.002	0.008	0.009
high	max underestimation [w]	5.255	3.501	3.318	0.206	0.193	0.230
	max overestimation [w]	1.315	2.446	2.425	0.952	2.451	2.847
	max rel. error [%]	45.8	41.9	40.2	7.3	22.0	26.1
	mean rel. error [%]	14.3	14.2	14.2	1.1	6.1	7.4
	1−correlation	0.060	0.045	0.044	0.001	0.010	0.012

mix, but not better. The correlation value, however, remains close to zero and indicates, as Fig. 4.7(a) does, that the general trend of the power distribution is captured. The results for experiments including shadowing are comparable to literature. *Dziong et al.* (1999) report their expected-coupling estimates to lie within 5–10 % of simulation; the mean error and correlation value that we obtained are comparable to the results of *Türke* (2006, Ch. 6).

To investigate whether an accurate computation of the expected coupling matrix including the effect of shadowing provides an estimation quality comparable to the no-shadowing cases, we additionally analyze the relation between simulation results and estimates based on the precise mean of the coupling matrix. Analytical methods for calculating this mean are beyond the scope of this thesis; some methods are cited above. For our investigation, we compute the mean coupling matrix with simulation.

Fig. 4.7(b) illustrates the quality of expected coupling estimates that use the accurate mean of the interference coupling matrix. The cells' load factor used for estimation is now calculated on the basis of this accurate mean. The dots in Fig. 4.7(b) hence have the same ordinate (observed power) as the ones in Fig. 4.7(a), but the abscissas are adjusted. This shifts the points back into position and the observed values lie close to a smoothed curve similar to the graphs in Fig. 4.5. The equivalent to Fig. 4.7 for grade of service estimates is given in Fig. B.2; they allow for similar conclusions. The right-hand part of Tab. 4.5 contains the same analysis for estimates derived with the precise average coupling matrix for all snapshot experiments conducted. The estimation quality is comparable to the estimates for the case of no shadowing presented above. The only qualitative exception is that there is both over- and underestimation

Table 4.6: Key results from conducted experiments in all settings: worst performance across traffic mixes and intensities in each case

	network	transmit power $\bar{p}^{\downarrow}$			grade of service $\lambda^{\downarrow}$ [a]		
		max rel. err. [%]	mean rel. err. [%]	1−corr.	max rel. err. [%]	mean rel. err. [%]	1−corr.
Lisbon	regular	24.1	4.8	0.008	3.8	0.5	0.013
	optimized	25.8	4.8	0.019	4.5	0.6	0.039
Berlin	regular	24.6	4.8	0.010	7.0	0.5	0.007
	optimized 1[b]	26.8	7.2	0.014	6.3	1.0	0.016
	optimized 2	25.4	6.7	0.012	6.2	1.1	0.005
	optimized 3	23.7	5.3	0.014	3.7	0.7	0.005

[a] Refined estimate
[b] Detailed results above

also for the low traffic case. The overall conservative bias of the estimates is retained. This experiment's conclusion is that an accurate computation of the expected coupling matrix is necessary for obtaining estimates that are precise for individual cells.

Experiments for other network configurations and scenarios. We briefly report on the results of comparable experiments for the remaining settings. The data for estimation quality of power and grade of service are presented in Tab. 4.6; for each network configuration we have listed the worst, *i. e.*, largest, maximum and mean relative error as well as the correlation complement across all service mixes and traffic intensities. The numbers demonstrate that the quality of estimates in the case that we discussed in detail above (Berlin optimized 1) is typical for all other cases investigated.

Other performance indicators related to capacity. From the performance indicators related to capacity, we have not yet discussed other-to-own-cell interference ratio and E_c/I_0 coverage. The other-to-own-cell interference ratio gives a general orientation on the quality of network planning, but it is of secondary importance. The quality of estimates of the other-to-own-cell interference ratio is depicted for the example network at medium traffic intensity in Fig. 4.8. For each cell, the figure plots the value estimated by the expected interference coupling method against the value observed in simulation. Because the values have a high dynamic range, logarithmic scales have been chosen. If no shadowing is considered (Fig. 4.8(a)), the estimation quality is acceptable, *i. e.*, all dots are close to the diagonal. There is a general bias towards underestimation. This tendency is more pronounced if shadowing is considered (Fig. 4.8(b)); this is aligned with the tendency to underestimation of power values in the medium and low traffic intensity cases. Even in the case of shadowing, however, estimated and observed values are highly correlated. Numerical results for the other-to-own-cell interference ratio estimation in the example configuration are listed in detail for all service mixes and traffic intensities in Tab. B.1 in the appendix.

Figure 4.8: Estimation quality for other-to-own-cell interference ratio
(downlink, network "Berlin optimized 1", medium traffic intensity, complete service mix)

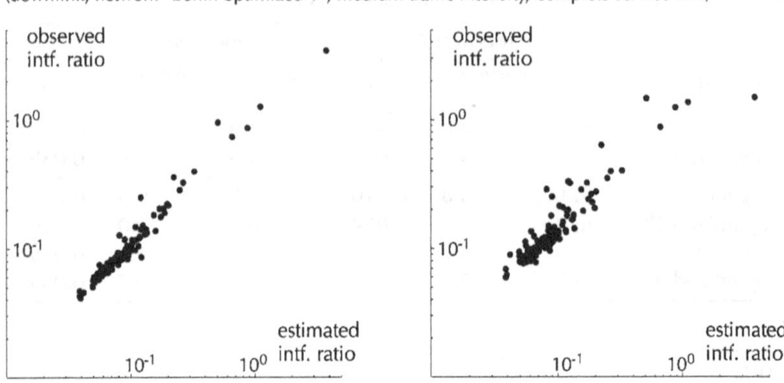

(a) No shadow fading (b) Including shadow fading

We eventually consider E_c/I_0 coverage and check whether critical areas are identified. We compare how the set of points with a coverage probability above 98 % can be approximated using the threshold-adjustment approach described in Sec. 4.2.2. Fig. 4.9 shows that an expected coupling analysis approximately identifies the areas with coverage problems. Fig. 4.9(a) plots the E_c/I_0 coverage probability as determined by simulation (including shadowing). The dark highlighted areas have more than 2 % outage probability. The E_c/I_0 level predicted by expected coupling is depicted in Fig. 4.9(b). Trials have shown that for a threshold adjustment of $\xi = 2.1\,\text{dB}$, membership in the set with a high coverage probability was correctly predicted in 93.4 % of the total planning area. As can be seen in Fig. 4.9, this enables a good indication of the areas in which the E_c/I_0-level needs to be increased for better coverage.

Moreover, trials over all Berlin network configurations with the different traffic mixes and service models have shown that the threshold need not be adapted to a specific setting. The detailed results are listed in Tab. B.2 in the appendix. While the optimal threshold adjustment varies from case to case, the coverage predictions do not lose precision considerably if a uniform threshold is picked. After an initial calibration, the expected-coverage method can thus be used efficiently.

4.5 Conclusions on system modeling and performance evaluation

To investigate the fitness of our system model and performance estimation schemes for use in optimization methods, we need to answer two questions: First, is the computational effort feasible? Second, does the information obtained enable the right planning choices? Computational efficiency is a hard constraint in optimization; high accuracy is desirable, if it can be obtained without excessive computation.

The efficiency of perfect load control and expected interference coupling apparently cannot be reduced significantly if equivalent information is to be generated. For

Figure 4.9: Approximating simulated E_c/I_0-coverage probability with expected coupling

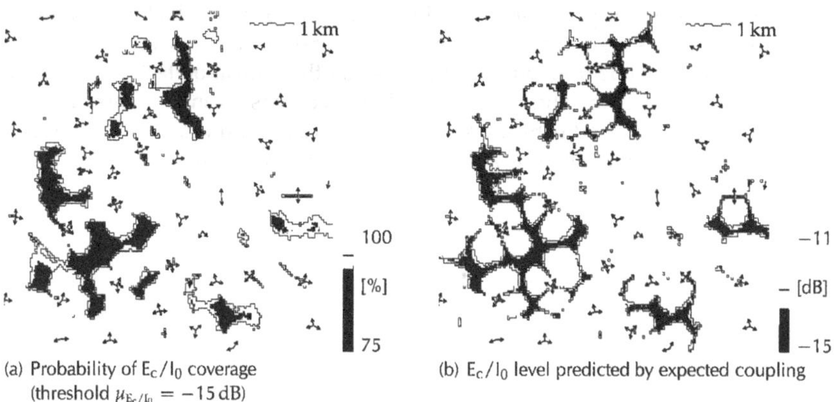

(a) Probability of E_c/I_0 coverage (threshold $\mu_{E_c/I_0} = -15\,\text{dB}$)

(b) E_c/I_0 level predicted by expected coupling

a typical representation of planning data, the expected coupling matrix (with the restrictions detailed above) can basically be calculated in time linear to the size of the input. This means that, asymptotically, no more time is needed than is necessary for reading the input once. In addition, we have shown how the estimates can be improved by using second order moments in an economical way. Solving a coupling complementarity system and obtaining the scaling factors in addition to power values is only as demanding as solving a traditional coupling equation system. This effort cannot be avoided if the coupling relations between cells are taken into account at all.

Validity of perfect load control. We have assessed the validity of perfect load control by comparing it to traditional, discrete load control schemes. We have shown that there is a discrete scheme for which the deviations from the idealized model in terms are small; the errors in estimating power and user blocking are on the order of magnitude of user granularity. We thus sacrifice the fine-granular view on individual users, but this causes no other major inaccuracies. Especially if put into perspective with standard simplifications taken in static modeling, the minor inaccuracies seem reasonable to obtain a top-level view reflecting the state of a network.

Quality of estimates for expected values of performance indicators. The accuracy of expected-coupling estimates has been researched by comparing them to the results of detailed Monte Carlo simulation. If no shadowing is considered, then the estimates match the results of simulation well for individual cells. The mean error is on the order of a few percent in all settings we investigated. In detail, the accuracy of estimates depends mainly on normalized traffic intensity and on the service mix. This is explained well by our simplified analysis: The more demanding users there are, the higher is the variance of traffic load, which decreases the precision. For low to medium traffic load, the estimates are comparatively precise; if many cells have high average load, then the estimates tend to have a pessimistic bias. In the case of shadowing, the precision for predicting capacity utilization for individual cells is mixed and the mean error is significantly increased. The overall picture, however, is grasped.

We can thus conclude from our experiments that the estimation techniques developed in this chapter are an efficient tool for taking decisions during optimization and for finding potentially good network designs. In any case, however, tentative solution candidates should eventually be evaluated by detailed simulation.

Things to remember: Evaluation of expected performance

- We introduce simplified Monte Carlo simulation as a reference method to determine expected network capacity and coverage.

- The expected traffic is aggregated over all services to the *normalized user load intensity* functions $T_\ell^{\uparrow}, T_\ell^{\downarrow} : A \to \mathbb{R}_+$. In the restricted random model (excluding shadowing) and under the best server assumption, the expected coupling elements can be efficiently calculated as integrals over the cell areas A_i, $i \in \mathcal{N}$:

$$\mathbb{E}(c_{ij}^{\uparrow}) = \int_{A_j} \frac{\gamma_i^{\uparrow}(x)}{\gamma_j^{\uparrow}(x)} T_\ell^{\uparrow}(x)\mathrm{d}x \qquad \mathbb{E}(p_i^{(\eta)}) = \int_{A_i} \frac{\eta^{\downarrow}(x)}{\gamma_i^{\downarrow}(x)} T_\ell^{\downarrow}(x)\mathrm{d}x$$

$$\mathbb{E}(c_{ii}^{\downarrow}) = \int_{A_i} \omega(x) T_\ell^{\downarrow}(x)\mathrm{d}x \qquad \mathbb{E}(c_{ij}^{\downarrow}) = \int_{A_i} \frac{\gamma_j^{\downarrow}(x)}{\gamma_i^{\downarrow}(x)} T_\ell^{\downarrow}(x)\mathrm{d}x$$

- We develop efficient deterministic estimates for expected network performance:
  - For *expected-coupling estimates*, we solve the complementarity system with the expected coupling matrix. Evaluating the performance indicators provides first-oder approximations of expected network performance.
  - A refined scheme for better approximating the grade of service includes second-order moments. The scheme requires one-dimensional numerical integration.

- Analysis of the pole equation shows:
  - The precision of expected-coupling estimates depends on the variance of the load factor, *i. e.*, the service mix and other-cell interference.
  - For networks operating in the typical load-range, power estimates are pessimistic and grade of service estimates are optimistic.

- Computational results for various scenarios, network configurations, and traffic levels show:
  - Perfect load control is a reasonable simplification.
  - The efficient estimates are satisfyingly accurate if shadowing is excluded.
  - With shadowing, the estimates are informative for taking planning decisions, but verification by simulation is required for tentative solution candidates.

Network performance optimization

We now strive to optimize a network design for best expected performance. The network properties can be modified by selecting site locations and manipulating static antenna and cell parameters. The performance indicators to be improved are usually grouped at top-level into cost, coverage, and capacity.

Capacity optimization is not yet satisfactorily addressed in scientific literature, because it requires a precise notion of interference coupling and the dependencies between cells. The classical static network model has been the only accurate tool available, so it is used as a black-box evaluation subroutine in heuristic search methods. These approaches, however, do not use the structure of the underlying problem. On the other hand, attempts at mimicking the classical static model in a transparent optimization formulation typically fail for complexity reasons. Practically feasible mathematical optimization schemes therefore basically include coarsely simplified capacity constraints into models that trade off cost versus coverage.

The system model developed in the previous chapters describes the effect of interference in a simple yet adept way; on its basis we can fill the gap with a new closed-form optimization model for network capacity. We formulate the objective in terms of the expected coupling matrix; the computational experiments in the previous chapter show that this allows optimizing expected network capacity in a deterministic setting. The distinguishing feature of our approach is the accurate description of coupling effects in a transparent model. We use the structure of the problem for developing new optimization approaches and deriving bounds on the optimum. Case studies on capacity optimization show that our approach is effective in large realistic settings, and the results allow insights into the complexity of capacity optimization.

This chapter is structured as follows: In Sec. 5.1, we review the general goals of performance optimization, we describe the available degrees of freedom, and we outline the relevant optimization methods. Sec. 5.2 surveys the related work in the area.

In Sec. 5.3, we specify objective functions for cost and coverage and our new optimization model for network capacity; we then combine the models for individual objectives to integrated models. We focus on downlink capacity optimization, but the approach can be extended to the uplink. In Sec. 5.4, we report on computational case studies in four realistic data scenarios and show how the new models and the improved understanding of network planning can be applied to real-world settings. Additional analyses on the outcomes of the case studies are presented in Sec. 5.5. We conclude in Sec. 5.6.

5.1 Prerequisites: objectives, parameters, and optimization methods

We describe the optimization problem at hand in detail by discussing the different objective functions and the parameters that can be manipulated. In addition, we provide a brief introduction to relevant mathematical optimization theory and methods. We use the notation introduced in the previous chapters, an overview of all symbols and their meaning is given on pp. 169ff.

Objective functions. There are three main objectives in radio network planning:

> "Any wireless network is a compromise between *capacity, coverage,* quality of service, and *cost.* Finding the best arrangement is a multidimensional, interdependent optimization problem." (*Symena Software & Consulting GmbH*, 2004, italics added)

Most software products share this multicriteria view of the problem and adopt the same (top-level) list of targets (e. g., *Actix Inc*, 2006; *Ericsson AB*, 2006, 2007; *Schema Ltd*, 2007). Optimizing the "quality of service" does not make sense in our setting, because each service defines a fixed quality requirement reflected in the uplink and downlink user load factors. With this exception, we adopt the view of trading off cost, coverage, and capacity. The specific implementation of the components, however, leaves room for interpretation. Let us comment on the possibilities for each performance metric and map them onto the performance indicators defined in Sec. 3.4 and listed in Tab. 3.1:

(a) *Coverage* denotes the ability of a user to connect to the network; it should be maximized. In the E_c sense, this has always been an objective for radio network planning. In the context of UMTS, the more involved requirement of E_c/I_0 coverage is added. Early stages of network planning tend to be driven by coverage; in some cases, there are coverage obligations (with respect to area or population) tied to spectrum licenses. The relevant performance indicators are the E_c-covered area $|A^{(E_c)}|$, the E_c/I_0-covered area $|A^{(E_c/I_0)}|$, the jointly covered area $|A^{(c)}|$, and their traffic-weighted versions $|A^{(E_c)}|^{\downarrow}_{\ell}$, $|A^{(E_c/I_0)}|^{\downarrow}_{\ell}$, $|A^{(c)}|^{\downarrow}_{\ell}$.

(b) *Capacity* denotes the amount of traffic that the system can serve simultaneously. In UMTS, capacity and coverage are linked through interference, and they are also conflicting: least interference is generated if cells have little overlap. For seamless

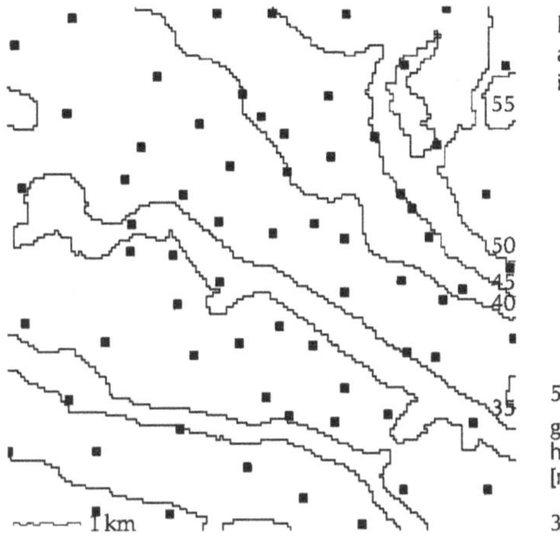

Figure 5.1: Potential site locations and terrain height profile in the Berlin scenario

coverage, on the other hand, overlap is needed. The performance indicators that address capacity are the average cell load $\bar{L}^{\downarrow}$, the grade of service $\bar{\lambda}^{\downarrow}$, and, as an indirect measure, the other-to-own-cell interference ratio $\bar{\iota}^{\downarrow}$. The grade of service is to be maximized, other-to-own-cell interference ratio and cell load are to be minimized.

(c) The *cost* at which services are provided and performance goals are attained has to be minimized. The deployment and operation of network infrastructure incurs cost, but also the commissioning of optimized settings into an operational network may be expensive. An example of a realistic cost model for antenna configuration changes is given in the white paper by *Symena Software & Consulting GmbH* (2003).

5.1.1 Degrees of freedom

The network parameters contained in Def. 1 on p. 26 and summarized on p. 33 are the indirect result of decisions on site location and configuration, antenna configuration, and pilot power level. Most commercial software tools offer the detailed planning parameters discussed in the following (e. g., *Actix Inc*, 2006; *Ericsson AB*, 2007; *Lustig et al.*, 2004; *Schema Ltd*, 2003*b*, 2007; *Symena Software & Consulting GmbH*, 2004); scientific authors usually pick some subset.

Site location and configuration. Site location and configuration are decided at an early stage of network planning, because they are tied to the largest expenditure and cannot easily be revised later on. In an already installed, mature radio network, sites are added or abandoned infrequently. The acquisition of new rooftops has become difficult, so mobile phone operators prefer to reuse existing sites, e. g., from a 2G radio network; this also reduces the cost of maintenance. The input to a site placement

Figure 5.2: Antenna diagrams and the effect of electrical downtilt (model Kathrein 742 212)

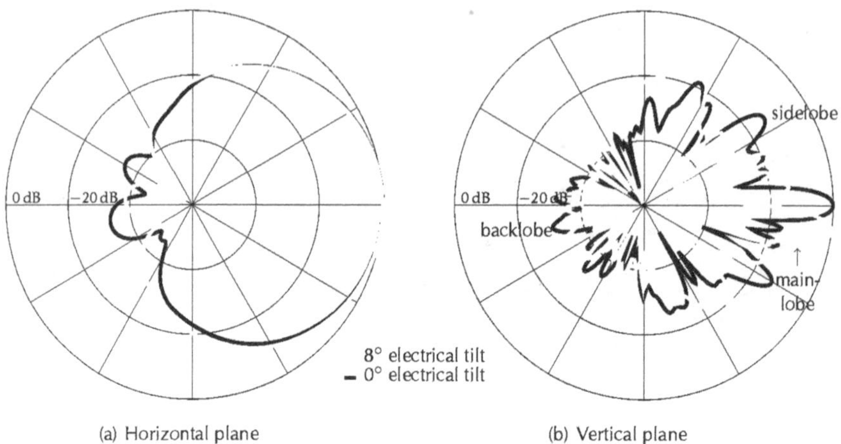

(a) Horizontal plane (b) Vertical plane

problem is hence a set of potential site locations; the candidates for Berlin, for example, are depicted in Fig. 5.1. *Hanly & Mathar* (2002) deduce a formula for optimum base station density in a simplified model; their main result is that the maximum distance between base stations is inversely proportional to the traffic intensity.

Parameters in site configuration include antenna height, base station and supporting hardware, and the number of sectors. The fundamental choice for antenna height is between antennas above or below rooftop level; the latter variant is sometimes chosen for urban microcells. Changing the height later on is avoided, because the structure at the site imposes hard constraints and a new building license might be required. The most expensive items to deploy are base station hardware and power amplifiers, the cost of all other equipment is small in comparison.

Antenna type. The characteristics of the *antenna model* significantly influence channel gains. Examples for 3G antennas can be found in the *Kathrein catalogue* (2001). The channel gain between an antenna and a given point depends on the point's directional position in the relative spherical coordinate system of the antenna. The channel gain is largest in the direction of the the antenna's main lobe. The ratio of this largest gain over the channel gain of an isotropic radiator at the same position is called the *antenna gain*. For directions off the main lobe, an offset is determined based on the *antenna diagram* specified by the manufacturer. An example diagram is shown in Fig. 5.2; it consists of the *horizontal* and the *vertical diagram* containing values measured along the equator and the prime meridian. The gain for all other angles is interpolated, for example using the methods of *Gil et al.* (2001) or *Scholz* (2000).

An antenna model's *horizontal* and *vertical opening angles* give an impression of the shape of the main lobe. The opening angle is defined as the largest angle containing

Figure 5.3: Channel gain functions for alternative antenna configurations

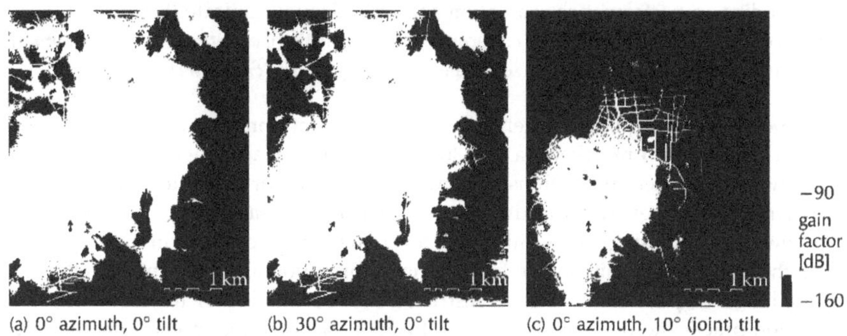

(a) 0° azimuth, 0° tilt (b) 30° azimuth, 0° tilt (c) 0° azimuth, 10° (joint) tilt

the main direction, in which attenuation is below 3 dB. Antennas with smaller opening angles are more focused. The vertical and horizontal opening angle of the antenna in Fig. 5.2 are 6.5° and 60°; the latter is a typical value for a three-sectorized site. A higher number of sectors calls for a smaller horizontal opening angle to avoid excessive overlap.

Antenna direction: azimuth and downtilt. The antenna direction is determined by the *azimuth* and *tilt* parameters, *i. e.*, the angles in the horizontal and vertical plane. The common coordinate system for measuring azimuth (from Arabic *as-samt*, "way, direction", *New Oxford Dictionary of English*, 1998) is inherited from astronomy: the point of reference is the north point of the horizon; angles are specified in degrees in the mathematically negative direction. An antenna pointing north thus is at 0° azimuth, east at 90° azimuth, and so on. Site-specific limitations on azimuth often have to be observed in practice, for example, due to physical features of the structure to which the antenna is attached.

Positive tilt values denote an orientation below the horizontal plane, so we speak of *downtilt;* the typical range is 2–15°. Higher values are used more frequently in areas with a high cell density. The antenna can be tilted mechanically or electrically with different consequences: Mechanical tilting rotates the antenna diagram in the vertical plane. Electrical tilting is implemented by modifying the relative phase of the different antenna elements. Thus, the antenna pattern itself is changed (*Nawrocki et al.*, 2006, Ch. 10.4.5). The difference is best pictured as tilting an umbrella (mechanical tilt) as opposed to closing it (electrical tilt). In practice, electrical tilt is usually preferred (*Niemelä & Lempiäinen*, 2004), as it reduces not only the impact of the antennas main lobe, but also of the side lobes (see Fig. 5.2(b)). In addition, mechanical tilt requires a manual adjustment, whereas electrical tilt can be adjusted remotely in modern systems. *Niemelä & Lempiäinen* (2004) investigate the impact of (uniform) mechanical tilt adjustment on network performance by simulation, the same is done by *Isotalo et al.* (2004) for electrical tilt.

Fig. 5.3 illustrates how tilt and azimuth variations influence signal propagation.[1] Many studies on performance evaluation and optimization indicate that tilt has more impact on network capacity and coverage than azimuth (*Gerdenitsch*, 2004; *Nawrocki et al.*, 2005; *Niemelä & Limpiäinen*, 2003; *Siomina*, 2007; *Türke & Koonert*, 2005).

Pilot power. The pilot power level has a large impact not only on coverage, but also on capacity because it represents a base load and scales the amount of interference. If pilot power is reduced, more transmit power is available for dedicated channels and less interference is generated in the network. *Türke & Koonert* (2005) have observed pilot power to be the single most important lever on network capacity in the downlink.

As the pilot is the main source for channel information and determines cell association, however, possible values are restricted. For soft handover to work well, it is a common practice to have pilot power uniform or varying in a limited interval only. If neighboring cells have different pilot settings, mobiles at the border might connect to the cell with higher link attenuation, thereby creating unnecessary interference. *Schema Ltd* (2003*b*) states a maximum dynamic range of 5 dB; *Türke & Koonert* (2005) allow a larger range but impose as a constraint that pilot power levels of neighboring cells differ by at most 1 dB. In general, high pilot values are used for obtaining coverage in rural areas, whereas in areas with high base station density, small values are favorable for interference reduction. The power on other common channels is usually set relative to the pilot channel, so the common channel powers $\boldsymbol{p}^{(\mathrm{C})}$ are determined by the pilot levels $\boldsymbol{p}^{(\mathrm{P})}$ and not considered a degree of freedom.

Combinatorial view. Radio network optimization is usually not modeled as a continuous, but as a combinatorial problem. The difference is that "in the continuous problems, we are generally looking for a set of real numbers [...]; in the combinatorial problems, we are looking for an object from a finite, or possibly countably infinite, set." (*Papadimitriou & Steiglitz*, 1982) Although some parameters are actually picked from a continuous range, there are two main reasons for the combinatorial view: First, network performance depends virtually discontinuously on the parameters. Second, it is practically impossible to adjust parameters with arbitrary precision. Some examples are: when adjusting the height of an antenna, the propagation properties change rapidly if the rooftop height is crossed; the electrical tilt of an antenna can only be adjusted by integer degree steps; a technician cannot set the azimuth with high precision without expensive equipment; the set of potential site locations is small and discrete. In the special case of pilot power optimization (which we do not pursue any further in this thesis), a continuous optimization with subsequent rounding is conceivable. For all other parameters, a limited set of potential configurations is usually input to optimization.

5.1.2 Combinatorial optimization methods in radio network planning

We give a brief overview of combinatorial optimization methods that are popular in radio network planning. This section is not complete; it should merely give a flavor

[1]Kathrein 742 265 antenna at 25 m in Lisbon's Barrio Alto.

of the techniques and algorithms that are mentioned throughout this chapter and provide references for the reader unfamiliar with the subject.

Scalar optimization problems. A scalar (*i. e.*, one-dimensional) optimization problem is specified by the *objective function* $f : \mathbb{R}^k \to \mathbb{R}$ and the *set of feasible solutions* $\boldsymbol{X} \subset \mathbb{R}^k$. We are interested in finding an element of $\boldsymbol{X}$ on which f assumes the smallest possible value. Formally, this is written as

$$\begin{aligned} \min \quad & f(x), \\ \text{s.t.} \quad & x \in \boldsymbol{X}. \end{aligned} \tag{SO}$$

The structure of specific problems depends on the properties of f and $\boldsymbol{X}$; in combinatorial optimization, $\boldsymbol{X}$ is usually finite or countable. Algorithms and techniques for solving this problem produce some feasible solution[2] and are commonly judged by two criteria:

(a) *Optimality* measures the quality of the produced solution. An element $x^* \in \boldsymbol{X}$ is called *optimal* for (SO) if $f(x) \geq f(x^*)$ for all $x \in \boldsymbol{X}$. If such an element exists, the number $f(x^*)$ is called the *optimum value* of (SO). Ideally, an optimization method produces a provably optimal solution or derives some bound on the optimum value, which is used to assess the quality of a found solution.

(b) *Efficiency* measures the time that the algorithm runs until it terminates. For theoretical analysis, it is popular to measure the efficiency of an algorithm in terms of the maximum number of elementary algorithmic steps needed for solving a problem instance of a given size. In practice, an algorithm is required to determine a solution within reasonable time using reasonable resources; what is "reasonable" depends on the circumstances.

We call methods that find the optimum or at least bound it *exact methods*. All other methods only find some (hopefully good) solution; they are called *heuristics* (from Greek εὑρίσκω, "to find", *New Oxford Dictionary of English*, 1998). Heuristics are important in practice because efficiency is often a high priority. For the types of problems considered in this thesis, a typical reasonable time frame is a few hours; a typical reasonable resource is a high-end personal computer.

Comprehensive overviews on exact methods for various combinatorial optimization problems are given by *Papadimitriou & Steiglitz* (1982) and *Schrijver* (2004). *Blum & Roli* (2003) provide a survey on heuristic techniques. Many successful combinatorial optimization algorithms are custom-crafted to specific problems and their structure. They are not easily adaptable because a slight and seemingly harmless change in the definition of $\boldsymbol{X}$ or f often destroys this structure. In the following, we focus on more general techniques used in radio network planning.

[2] This is of course impossible if $\boldsymbol{X}$ is empty. We will tacitly assume that this is not the case and that the *feasibility problem* of finding *any* element in $\boldsymbol{X}$ is easy. In network planning, it is often trivial.

Exact methods. *Exhaustive search* or *enumeration* consists in checking the value of f for all elements in the solution space $\boldsymbol{X}$. This simple technique can be applied successfully if f can be evaluated efficiently and $\boldsymbol{X}$ is easy to enumerate and small, or if large parts can be excluded. The best solution found is obviously optimal.

A special case on the border between continuous and combinatorial optimization are linear optimization problems or *linear programs* (*Chvátal*, 1983). Formally, we consider the case that f and $\boldsymbol{X}$ have a rational linear description, *i.e.*, $f(\boldsymbol{x}) = \boldsymbol{c}^t\boldsymbol{x}$ and $\boldsymbol{X} = \{\boldsymbol{x} \in \mathbb{Q}^k | A\boldsymbol{x} \leq \boldsymbol{b}\}$ for some $\boldsymbol{c} \in \mathbb{Q}^k$, $\ell > 0$, $A \in \mathbb{Q}^{\ell \times k}$, and $\boldsymbol{b} \in \mathbb{Q}^\ell$. An optimal solution of a linear program can be found efficiently by interior point methods based on the algorithm of *Karmarkar* (1984) and, in practice, by the *simplex algorithm*. A generalization of linear problems that can also often be treated successfully are *convex* optimization problems.

Feasible solutions are often required to be integral in combinatorial optimization. For example, the number of antennas in a network can only be integral, or binary variables corresponding to a yes/no decision may assume only 0 or 1 as a value. If the structure of the problem remains linear, we speak of an *integer (linear) programs* (IP, *Schrijver*, 1986). The additional constraint that the solution $\boldsymbol{x}$ be integral is formally written as $\boldsymbol{X} = \{\boldsymbol{x} \in \mathbb{Z}^k \mid A\boldsymbol{x} \leq \boldsymbol{b}\}$. A wide range of practical problems can be formulated in the form of an integer program (*Nemhauser & Wolsey*, 1988). *Mixed integer linear programs* (MIP) generalize IPs in that only a part of the variables must be integral and the others may be continuous.

The most efficient general solution methods for MIPs use a branch-and-bound-and-cut scheme, which employs the principle of exhaustive search but exploits the structure of the problem to avoid complete enumeration wherever possible. Bounds on the optimum value are determined by linear programming. In general, a short running time of (mixed) integer programming algorithms is not guaranteed. Whether or not integer programming methods are an adequate tool for a specific problem depends on the model formulation and problem size; in many practical cases these methods are successful. An overview and access to many software solvers is provided by the NEOS server (*Dolan et al.*, 2002). There are also methods for nonlinear integer programming (*Li & Sun*, 2006), but so far most large nonlinear problems remain intractable.

Search- and population-based metaheuristics. A *metaheuristic* is "a set of algorithmic concepts that can be used to define heuristic methods applicable to a wide set of different problems" (*Doriga & Stützle*, 2004). *Search heuristics* start with some initial feasible solution and then iteratively move to other solutions, depending on properties of the current solution and possibly the search history. Often the next solution candidate is selected from some *neighborhood* set $N(\boldsymbol{x}) \subset \boldsymbol{X}$ defined for the current solution $\boldsymbol{x}$. The simplest search heuristic, *local search,* starts with an initial solution and iteratively moves to the best candidate within the current neighborhood until no further improvement can be achieved. The search hence always terminates with a local optimum, which is not necessarily globally optimal.

Some methods generalize local search and introduce features to overcome its shortcomings, specifically to escape from suboptimal local minima. Probably the oldest search algorithm with an explicit strategy for doing so is the randomized method of *simulated annealing;* a move worsening the objective function may be executed with a probability that is slowly reduced during execution to obtain convergence. Under certain circumstances, convergence to an optimal solution (with high probability) can be guaranteed. Another popular scheme is *tabu search:* The recent history of moves is remembered for some time and returning to an already discovered solution or performing several subsequent similar moves is forbidden (tabu) for a specified time.

Some metaheuristics deal with a set of solutions (a population) in every iteration and apply operators to manipulate this set. *Genetic algorithms* and *evolutionary algorithms* mimic biological evolution: moves are made from members of the population (mutation), different members are combined (recombination/reproduction), and the worst members of the population are discarded (selection).

The merits of metaheuristics are subject to debate. Their effectiveness depends critically on a useful implementation of the generic concepts of neighborhood or mutation operators. A robust, problem-specific heuristic is often more useful than a generic metaheuristic scheme; *Wolpert & Macready* (1997) theoretically substantiate this claim. On the other hand, there are problems for which seemingly naive implementations of metaheuristics outperform other methods. In radio network planning, metaheuristics are popular; for example, a software product by *Optimi, Inc* (2006*a*) uses evolutionary algorithms, simulated annealing, and tabu-search; *Schema Ltd* (2003*a*,*b*) uses evolutionary and genetic optimization.

5.1.3 Multicriteria optimization

Given a vector-valued objective function $f : \mathbb{R}^k \to \mathbb{R}^\ell$ and a set $X \subset \mathbb{R}^k$, the problem

$$\begin{aligned} \min \quad & f(x)\,, \\ \text{s.t.} \quad & x \in X \end{aligned} \tag{MO}$$

is called a *multi-objective* or *multicriteria* optimization problem. Terminology and notation are diverse; we stick to the monograph by *Ehrgott* (2005), which covers most theoretical aspects. *Osyczka* (1984) gives an overview of pragmatic approaches in engineering.

The definition of the minimum in (MO) is a generalization of the one-dimensional case; it uses a partial order in the objective space $\mathbb{R}^\ell$, most often the component-wise order "$\leq$" as defined on p. 25. An *optimal* or *nondominated solution* is a point $x^* \in X$ for which there is no other point $x \in X$ with $f(x) \leq f(x^*)$. Unlike in the single-objective case, we cannot characterize x^* by demanding that $f(x^*) \leq f(x)$ for all points $x \in X$, because vectors need not be comparable and such a point does not exist in general. In other words, a solution is optimal if no component of the objective function can be improved without increasing the value of another one.

The *Pareto front* of (MO) describes the trade-off between the objectives; it is defined as the set of all optimal solutions. The Pareto front for a two-dimensional optimization problem is illustrated in Fig. 5.4; the feasible set (in objective space) is the set of all possible objective vectors, *i.e.*, the image of X under f. It is obvious that points

Figure 5.4: Multicriteria optimization—Pareto front and weighted sum scalarization

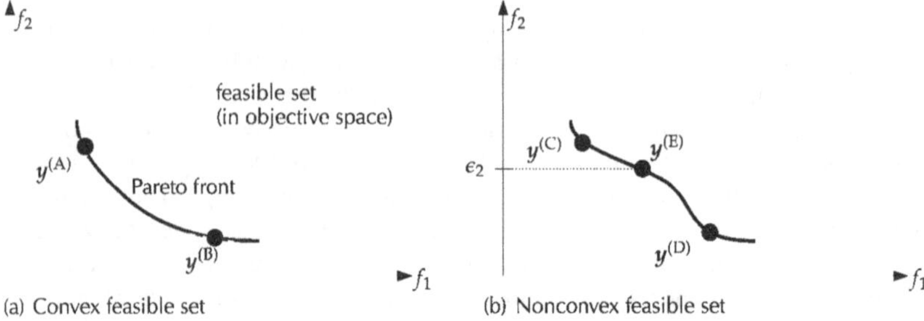

(a) Convex feasible set (b) Nonconvex feasible set

$y = f(x)$ contained in the Pareto front are the most interesting solutions, but these points may have very different characteristics.

Scalarization methods. Scalarization of multi-objective problems, *i. e.*, their transformation into a single-objective problem, is a popular tool to find points in the Pareto front; *weighted-sum scalarization* is probably most often used. A vector of nonnegative weights $w \in \mathbb{R}^{\ell}_{+}$ for the components of the objective function is selected and the following single-objective problem is solved:

$$\begin{aligned} \min \quad & w^t f(x) \\ \text{s.t.} \quad & x \in X \end{aligned} \tag{SO(w)}$$

The optimal solution is always a member of the Pareto front of (MO). The points $y^{(A)}$ and $y^{(B)}$ in Fig. 5.4(a) correspond to the optimal solution values under different weight vectors indicated by the dashed lines. Some software products for radio network planning use weighted-sum scalarization and let the user pick suitable weights (*Optimi, Inc*, 2006a; *Symena Software & Consulting GmbH*, 2004). The weighted-sum method often has a special advantage: If the optimization problem for all individual objective functions can be solved efficiently, this holds also for the scalarized problem.

The drawbacks of the weighted-sum approach are the difficulty of choosing a weight vector and that in general not all points in the Pareto front can be produced. For picking a weight vector such that the solution to (SO(w)) represents an appropriate compromise, information (or intuition) on the structure of the Pareto front is required. Furthermore, a comparison between quantities deemed incomparable in the first place is artificially defined. Moreover, desirable compromise solutions may be impossible to find with weighted sum scalarization. An example is depicted in Fig. 5.4(b): no point in the segment of the Pareto front joining the points $y^{(C)}$ and $y^{(D)}$ is the optimum solution to any problem of the type (SO(w)).

An alternative scalarization approach is the *ϵ-constraint* method first introduced by *Haimes et al.* (1971). It focuses on one objective function component with index i^*

and requires the other components of f to stay within certain bounds $\epsilon \in \mathbb{R}^\ell$:

$$\begin{aligned} \min \quad & f_{i^*}(x) \\ \text{s.t.} \quad & f(x) \leq \epsilon \\ & x \in X \end{aligned} \tag{SO(i^*,ϵ)}$$

This overcomes the above theoretical problem of the weighted-sum approach: any point on the Pareto front is the solution to a problem of type (SO(i^*,ϵ)) for a suitable choice of ϵ. The point $y^{(E)}$ in Fig. 5.4(b), for example, is the optimum solution for $i^* = 1$ and the indicated choice of ϵ_2. It is also attractive in practice, because an isolated understanding of the single components suffices for fixing ϵ; different experts may decide the different components. On the downside, however, even if optimization for any of the components is easy, their combination with the ϵ-constraint method is (provably) more difficult for many problems.

Hybrid scalarization combines the weighted-sum and ϵ-constraint approaches: a weighted sum of the objectives is optimized while individual bounds are imposed on the components. The *value-function method* generalizes weighted-sum optimization by admitting an arbitrary aggregation of the objectives to a single number.

Hierarchical and sequential optimization. In *hierarchical optimization* (introduced by *Walz*, 1967), a priority order of the objectives is set. The functions are optimized sequentially, and in each step the values of the previous objectives are only allowed to deviate by a certain margin from the respective optimum. We solve the problems

$$\begin{aligned} \min \quad & f_i(x)\,, \\ \text{s.t.} \quad & f_k(x) \leq (1+\epsilon_k)\min(\text{SO}(k)) \qquad \text{for all } k = 1,\dots,i-1\,, \\ & x \in X\,. \end{aligned} \tag{SO(i)}$$

for $i = 1,\dots,\ell$. If $\epsilon = 0$, we speak of *sequential optimization*, which corresponds to optimization under the lexicographic order on $\mathbb{R}^\ell$. Sequential optimization is used if the objective functions are loosely coupled or to decompose a complex problem into a sequence of manageable problems. In wireless network planning, it is, for example, popular for coverage optimization with subsequent channel assignment in GSM (*Eisenblätter*, 2001) or WLAN networks (*Hills*, 2001).

Multicriteria combinatorial optimization and metaheuristics. In many cases, the techniques listed for single-objective optimization are used for scalarized versions of multi-objective combinatorial problems. Besides, there are multicriteria extensions of some algorithms, for example, of the simplex method (*Ehrgott*, 2005) and of most metaheuristics (*Ehrgott & Gandibleux*, 2002). Population-based metaheuristics are particularly amenable to this extension as they already operate on multiple solutions. In their multicriteria form, evolutionary and genetic algorithms explicitly aim at finding a set of solutions that approximates the entire Pareto front. Concepts of diversity, gender, and special distance measures are used to this end (*Coello*, 2000; *Deb*, 2001).

5.2 Survey of network planning literature

We will first provide a brief overview on the different lines of research in the field of UMTS radio network planning in Sec. 5.2.1. In Sec. 5.2.2, we classify the literature according to the choices made in four central modeling questions. We also discuss the implications of the different options as to their impact on model complexity and accuracy. The current work is related to previous contributions in Sec. 5.2.3. There is no universally accepted radio network planning benchmark in the scientific community, so the comparison can only be made at a conceptual level.

5.2.1 Overview

We group contributions into three major categories according to how they relate to the classical static system model introduced in Sec. 2.4. The individual contributions are grouped by authors and listed in no particular order.

General, legacy, and simplified models. Generic radio network planning models and models targeting previous technologies do not use the static model. They hence have no notion of interference coupling or signal quality.

Hurley (2002) considers generic site location and antenna configuration planning (primarily targeting GSM). He specifies coverage, cost, capacity, handover area, and overlap as potential objective functions. For the problem of optimizing a weighted sum of these objectives, different start heuristics and a simulated annealing scheme with several problem-specific neighborhood definitions are proposed and tested.

Jedidi et al. (2003) argue that automated optimization methods are little accepted by radio network planning practitioners because resulting cell geometries are awkward. Some models include geometrical notions to address this problem. *Reininger & Caminada* (2001) introduce the constraint that cells be connected into a site positioning and configuration model. Full signal-level coverage of test points is also required, and cost, traffic, and overlap are the objectives. The idea is subsequently refined by adding a cell shape factor as an objective. Cells with a small shape factor resemble a circle, so the objective is to be minimized in order to come closer to this ideal. This has been handled as a constraint (*Jedidi et al.*, 2003) and as a second objective besides overlap (*Jedidi et al.*, 2004) in a multi-objective evolutionary algorithm.

Galota et al. (2001) develop an approximation algorithm for a base station positioning problem in which they penalize approximated interference coupling. The authors specify a polynomial time approximation scheme for the problem and furthermore show that (for Euclidean attenuation) their problem is strongly NP-hard and does not admit a fully polynomial time approximation scheme. *Glaßer et al.* (2005) develop similar approximation schemes for maximizing the coverage achieved with a fixed number of cells (maximum covering location problem) and minimizing the number of cells for complete coverage (minimum set covering problem).

Tutschku & Tran-Gia (1998) specify an algorithm for generating a discrete set of *demand points* representing a given spatial traffic distribution. *Tutschku* (1999) then describes optimization models for adjusting base station position and cell radius (*i. e.*, pilot power). In the simple version, signal-level coverage and network cost are considered as objectives. Greedy heuristics and simulated annealing solve maximum covering location and minimum set covering problems. Additionally, constraints on co-channel interference are added in a refined version of the optimization model. For W-CDMA network planning, *Leibnitz* (2003) adds the squared deviation from a cell-overlap target value to the objective function and solves the model with a greedy, set-covering-inspired heuristic followed by a fine-tuning phase based on geometric considerations.

Mathar & Niessen (2000) formulate and test optimization models for base station positioning with coverage, overlap, and interference as objectives.

The static model with decision variables. A second class of contributions formulate a deterministic optimization model, which is based on the full static model with additional decision variables for base station positioning and antenna configuration. *Eisenblätter et al.* (2002) describe the most detailed version; the works of *Amaldi et al.* (2006, 2003) and *Schmeink* (2005) also belong into this category. All cited publications use the number of served users as an objective (considering E_c coverage, base station power constraints, and partly E_c/I_0 coverage). The complete model, however, is too complex to be solved for practical instances, so different relaxations and simplifications are proposed. We shall give details on the different approaches.

The model described in *Eisenblätter et al.* (2002) contains binary variables for site positioning, configurations, and assignment of mobiles to servers in several (independent) snapshots. In addition, continuous power variables are included and related through CIR inequalities. *Eisenblätter et al.* (2003) approximate the problem by a decoupled set-covering problem to minimize a weighted sum of up- and downlink cell powers and cost under a full coverage constraint. In a later contribution (*Eisenblätter et al.*, 2004) propose solving smaller instances of the MIP for subproblems.

Amaldi et al. (2003) treat base station locationing and antenna configuration with a similar model including CIR inequalities. The degrees of freedom in their experiments are azimuth, tilt, and antenna height. They derive a simplified model for "power-based power control," assuming that power control aims at maintaining a particular signal level (rather than signal quality); this makes power variables obsolete. They furthermore derive an upper bound on the number of connections per sector by calculating with a fixed other-to-own-cell interference ratio. MIP solver software (for small instances), randomized greedy heuristics, and tabu search minimize a weighted sum of site cost and downlink powers for achieving full coverage (*Amaldi et al.*, 2003). In a later contribution (*Amaldi et al.*, 2006), a weighted sum of the number of served users and installation cost is the objective. The simplified model of power control can reportedly be evaluated faster than the accurate model based on signal quality.

Mathar (2001) specifies an optimization model for maximizing joint E_c and E_c/I_0 coverage based on the demand-node model of *Mathar & Niessen* (2000), which is extended by signal quality constraints. *Mathar & Schmeink* (2001), in addition, constrain the number of users per cell and add a notion of soft handover. The basic model is solved on small instances with special branch-and-bound methods featuring a greedy branching step. *Schmeink* (2005) details the full model including CIR constraints, but then uses a "pilot-based" model for optimization: no individual link powers are decided, but cell powers are set to fixed (parameterized) levels.

The line of work started by *Värbrandt & Yuan* (2003) focuses on adjusting pilot power values. It is assumed that all cells transmit at maximum power, and the pilot power level is adjusted according to the E_c/I_0 level computed under this assumption. For small instances and complete coverage, the model is solved with a MIP solver (*Värbrandt & Yuan*, 2003). A model with parameterized level of coverage is solved by a heuristic method based on Lagrangian relaxation (*Siomina & Yuan*, 2004). *Siomina & Yuan* (2005) then add minimum requirements on the size of soft handover area, which is ensured by a post-processing step on heuristically found solutions to the previous model. In the latest contribution (*Siomina et al.*, 2006), antenna configuration is added as a degree of freedom and simulated annealing is the method of choice.

Melachrinoudis & Rosyidi (2001) design (pre-UMTS) CDMA networks. They manipulate base station power, antenna height, and location for optimizing a weighted sum of network cost, signal quality, and estimated blocking rate.

The static model as an evaluation oracle. Eventually, some contributions, notably by commercial players, do not disclose their performance evaluation model completely, but use it in search procedures or other metaheuristics. These contributions can be seen as depending on an evaluation oracle. Evaluation is presumably based on the static model, because W-CDMA is explicitly addressed, but it does not involve (full) simulation for performance reasons.

Murphy et al. (2005) optimize azimuth and tilt such as to reduce load, avoid code exhaustion, and reduce link power. The power is estimated in an expected-coupling fashion.[3] The optimization model is solved by local search. *Weal et al.* (2006) build upon the same evaluation model and use different metaheuristics for optimizing a weighted sum of power consumption, coverage, pilot pollution, and others.

Türke & Koonert (2005) first use geometric heuristics for quickly finding reasonable values for pilot power and antenna configuration and then improve on the start solution by search methods, notably local search, simulated annealing, and tabu search. The performance of alternative configurations is evaluated by an expected-coupling scheme similar to our expected coupling estimates. The precise objective function as well as the neighborhood definition are not detailed.

Gerdenitsch et al. (2003) adjust pilot power and antenna tilt of an existing radio network. One snapshot is used for optimization; it is extended as more users can be served. As a solution method, the authors propose a simulated annealing heuristic with a set of rules for modifying the current configuration based on load; the rules are modified later (*Gerdenitsch et al.*, 2004). *Gerdenitsch* (2004) complements the previous methods with a genetic algorithm optimizing a weighted sum of the grade of service, coverage, and soft handover area and adds geometry-inspired heuristics.

Altman et al. (2002) introduce the automatic cell planner *Oasys*, which employs a genetic algorithm to manipulate common channel powers and antenna configura-

[3] This is not detailed in the paper but was confirmed by the first author in private communication.

tion. No concrete information on the algorithm, objectives, and evaluation is given. However, as "the evaluation time of a network with a few hundreds of sectors is on the order of a second" (*Jamaa et al.*, 2004), it is unlikely that simulation is used as a subroutine; traffic intensity is reportedly considered, nonetheless. *Jamaa et al.* (2003) report that automated optimization produces better results than manual optimization in most cases. The approach is extended to multi-dimensional optimization of capacity and coverage later (*Jamaa et al.*, 2004). *Jamaa et al.* (2005) then proceed to optimizing only a subset of sectors in considerably shorter time; the selection of sectors to consider is based upon an "interference matrix" (different from our coupling matrix).

5.2.2 Classification of approaches

We now classify the cited works according to four categories: modeling of power control and interference; choice of objectives and treatment of multiple objectives; modeling of spatial user distribution; and solution algorithm. Tab. 5.1 summarizes the main options explored so far in these four dimensions and lists our choices.

Power control and interference. The modeling of power control and of the coupling of coverage and capacity through interference is crucial. The most accurate account of W-CDMA interference coupling is contained in the complete, explicit models, which contain individual link powers, CIR inequalities, limits on (downlink) cell powers, and E_c/I_0 coverage constraints (*Amaldi et al.*, 2006; *Eisenblätter et al.*, 2002; *Schmeink*, 2005). The same might hold true also in the contributions using an evaluation oracle (*Gerdenitsch*, 2004; *Gerdenitsch et al.*, 2004, 2003; *Türke & Koonert*, 2005). The model used by *Murphy et al.* (2005) and *Weal et al.* (2006) seems to account for complete interference coupling, but it apparently does not contain capacity limits per cell.

The most popular simplification of interference coupling is to focus on the downlink and calculate with fixed cell power levels; this effectively eliminates interference coupling. The fixed value may be a common pilot power level (*Mathar*, 2001; *Mathar & Schmeink*, 2001), an interference-power value individually parameterized per cell (*Schmeink*, 2005), or the maximum feasible transmit power (*Siomina et al.*, 2006; *Siomina & Yuan*, 2004, 2005; *Värbrandt & Yuan*, 2003). In these cases, E_c/I_0 coverage is checked easily. *Amaldi et al.* (2006, 2003), in contrast, use a pure uplink model for calculations; their "power-based power control" scheme corresponds to a linearization of interference coupling.

Contributions addressing previous technologies like GSM (*Hurley*, 2002; *Mathar & Niessen*, 2000), geometry-inspired models (*Jedidi et al.*, 2004; *Reininger & Caminada*, 2001), and some earlier work targeting UMTS (*Galota et al.*, 2001; *Leibnitz*, 2003) does not consider power control at all.

Multiple objectives. Pure modeling papers without computational experiments occasionally list a number of objectives (e. g., *Eisenblätter et al.*, 2002; *Reininger & Caminada*, 2001). A few contributions use surrogate objective functions outside of the cost-coverage-capacity scheme: A multicriteria algorithm is used by *Jedidi et al.* (2004); *Melachrinoudis & Rosyidi* (2001) optimize a weighted sum of grade of service, (mini-

Table 5.1: Modeling and optimization of UMTS radio networks

	previous literature	this work
spatial user distribution	disregarded snapshot(s) – random – representative mean of distribution	mean of distribution
multiple objectives	cost vs. coverage – ϵ-constraint (minimum set covering/ max coverage locationing) – weighted sum coverage and capacity – ϵ-constraint – weighted sum – multiple solutions	multicriteria framework ϵ-constraint scalarization for – coverage – cost – capacity
power control and interference	disregarded overlap fixed transmit powers interference coupling – linearized – fully nonlinear	full coupling approximations – linear – convex
solution method	general heuristics – search heuristics – metaheuristics specialized heuristics – geometric – custom neighbor-hood/operators exact – approximation algorithm – MIP solver – special branch-and-bound	MIP solver search heuristics – 1-opt – k-opt

mum) call quality, and total cost. Virtually all other contributions use scalarization methods on coverage, capacity, and cost objectives.

There are two main classes: those including site selection and those focusing on capacity optimization. The works considering site placement attach a cost value to each site and handle the total cost of a network design as an objective. The optimization then primarily explores the tradeoff between coverage and cost, the precise meaning of "coverage" depending on the underlying model of interference; an additional approximation of capacity and specifics of UMTS is sometimes added. There are some weighted-sum approaches (*Amaldi et al.*, 2006; *Galota et al.*, 2001; *Hurley*, 2002), the remaining contributions use the ϵ-constraint method for scalarization. The problems solved are hence variants of either the maximum coverage location problem (if cost is constrained and coverage maximized) or minimum set covering problem (vice versa). Both forms are attacked by *Tutschku* (1999) for E_c coverage and by *Glaßer et al.* (2005) for E_c/I_0 coverage. Set covering is solved by *Amaldi et al.* (2003) and *Leibnitz* (2003), the latter complemented by an additional additive overlap adjustment term. *Mathar* (2001), *Mathar & Schmeink* (2001), and *Schmeink* (2005) solve covering location problems.

All works including site selection simplify interference coupling, so cell load or capacity can only be included to a limited degree. Under the assumption of fixed transmit powers, the individual link powers can be precomputed and used in a knapsack capacity condition per cell (*Hurley*, 2002; *Mathar*, 2001; *Mathar & Schmeink*, 2001; *Schmeink*, 2005). The only optimization model in this category with a notion of transmit powers (*Amaldi et al.*, 2003) uses them as a second, weighted objective.

Works focusing on cell configuration do not consider cost, but maximize cell capacity and coverage. This is addressed with a multi-objective genetic algorithm by *Jamaa et al.* (2004, 2005). *Gerdenitsch et al.* (2004, 2003) maximize coverage with capacity constraints. For the genetic algorithm in *Gerdenitsch* (2004), a weighted sum of grade of service, coverage, and overlap (soft handover area) is used as the fitness function. *Weal et al.* (2006) optimize for a weighted sum of transmit power, coverage, grade of service, pilot pollution, and pilot power harmonization. The contributions by *Värbrandt & Yuan* (2003), *Siomina & Yuan* (2004, 2005), and *Siomina et al.* (2006) constrain coverage and maximize capacity by minimizing pilot powers.

The required level of coverage can make a difference in complexity: A 100 % coverage constraint renders binary coverage decision variables (specifying which users to cover) obsolete (*Amaldi et al.*, 2003; *Siomina et al.*, 2006; *Värbrandt & Yuan*, 2003). Few contributions explicitly investigate the tradeoff involved; only *Siomina & Yuan* (2004) determine capacity for varying requested coverage levels.

Random influences: spatial user distribution. Contributions usually consider the spatial user distribution, unless transmit powers are assumed fixed and full coverage is required (*Siomina et al.*, 2006; *Värbrandt & Yuan*, 2003), or if geometric criteria are emphasized (*Jedidi et al.*, 2004).

Three directions can be discerned: random snapshots, special representative snapshots, and mean values of the distribution. *Gerdenitsch et al.* (2004, 2003) and *Ger-*

denitsch (2004) use a single, randomly drawn snapshot. Several randomly drawn snapshots are used by *Eisenblätter et al.* (2003, 2004, 2002); their precise number is reportedly low. The remaining approaches consider traffic nodes, traffic points, service test points, or demand points instead of users; this hints at a representative configurations. In particular, the demand-node concept of *Tutschku & Tran-Gia* (1998) can be viewed as a representative snapshot. It is used by *Leibnitz* (2003), *Galota et al.* (2001), *Glaßer et al.* (2005), and *Mathar & Niessen* (2000). Weights are added to the demand-nodes by (*Amaldi et al.*, 2006, 2003), *Mathar* (2001), *Mathar & Schmeink* (2001), *Reininger & Caminada* (2001), and *Hurley* (2002). *Siomina & Yuan* (2004, 2005) use mean values of the traffic distribution to determine weighted coverage. The mean values are used to approximate average transmit power levels by *Murphy et al.* (2005); *Türke & Koonert* (2005); *Weal et al.* (2006).

If weights are attached to users, snapshot-based optimization is equivalent to calculating on mean values because the "average snapshot" can be considered by placing users on a regular grid with weights corresponding to the traffic intensity. As the size of a snapshot-based optimization problem increases with the number of users, however, this approach does not scale well.

Solution methods. The only exact algorithms employed in realistic settings are integer programming techniques for simplified linearized models; approximation algorithms are only used in theory (*Galota et al.*, 2001; *Glaßer et al.*, 2005). MIPs representing small instances or small subproblems are mostly solved by generic software solvers (*Amaldi et al.*, 2006, 2003; *Eisenblätter et al.*, 2004; *Värbrandt & Yuan*, 2003), sometimes terminating computations before reaching optimality (*Mathar & Niessen*, 2000). Numerical difficulties are reported even for small cases (e. g., *Amaldi et al.*, 2006). *Mathar* (2001), *Mathar & Schmeink* (2001), and *Schmeink* (2005) employ a custom branch-and-bound scheme. Occasionally, integer programming is used to solve heuristic approximations (*Eisenblätter et al.*, 2003).

The range of viable heuristic solution algorithms depends largely on the computational cost of evaluating the objective function. The fastest optimization methods do not determine cell powers and interference levels at all; an example is the geometry-inspired heuristics for site configuration by *Türke & Koonert* (2005); *Gerdenitsch* (2004) uses only five network evaluations for his geometry-inspired "analytic" optimization. For problems involving site-positioning, reasonable start solutions are often found with specialized greedy set-covering heuristics (*Amaldi et al.*, 2006; *Leibnitz*, 2003; *Tutschku*, 1999).

For optimization considering full interference-coupling in large instances of simplified models, search heuristics are popular. All neighborhood definitions described for sector configuration correspond to 1-opt schemes, *i. e.*, only one antenna configuration is manipulated at a time. Straightforward local search is rare (*Murphy et al.*, 2005); simulated annealing is the most popular variant (*Gerdenitsch et al.*, 2004, 2003; *Hurley*, 2002; *Melachrinoudis & Rosyidi*, 2001; *Siomina et al.*, 2006; *Tutschku*, 1999; *Türke & Koonert*, 2005); tabu search is chosen only by *Amaldi et al.* (2006, 2003).

The most time-consuming heuristics applied are general purpose metaheuristics;

notably genetic and evolutionary algorithms have been proposed. They are used by *Gerdenitsch* (2004) and *Weal et al.* (2006), and for multi-objective optimization by *Jedidi et al.* (2004) and *Jamaa et al.* (2004, 2003, 2005). *Gerdenitsch* (2004) reports 150 000 network evaluations during a run of the genetic algorithm. A specialized heuristics is the Lagrangian scheme developed by *Siomina & Yuan* (2004).

5.2.3 Positioning of the current work

The modeling options that we explore are listed in Tab. 5.1. The current work differs from previous mathematical contributions in that we take into account the full information on traffic distribution and use an accurate model of interference coupling. Beyond the previous engineering approaches, we formulate a transparent optimization model that allows to use the structure of the capacity optimization problem; for example, we derive lower bounds and new linear and convex approximations of nonlinear interference coupling. In addition, we embed capacity optimization into a multicriteria framework; this highlights the trade-offs involved and makes the new developments adaptable to practical situations. Furthermore, our methods are tested on four realistic datasets; their different characteristics allow additional insights, and our computational part is more extensive than any other public study on capacity optimization.

5.3 Optimization models

We shall now specify optimization models for network planning. Our strategy is to specify individual objective functions for cost, coverage, and capacity and to combine these subsequently with methods from multicriteria optimization. The models for cost and coverage are known, the model for capacity is a new contribution. We strive at keeping each component simple and independent from the other ones. In particular, we only use deterministic models based on the estimations developed in Ch. 4 using medians of attenuation and expected coupling.

We treat cost and coverage in Sec. 5.3.1 In Sec. 5.3.2, we present and discuss our new capacity optimization model. The objective function is nonlinear and even nonconvex; we therefore approximate it with a mixed integer programming model, which is specified in Sec. 5.3.3. The combined optimization models are then developed in Sec. 5.3.4. The notation introduced in the previous chapters is used in the following; a complete list of symbols with references to their introduction is given on pp. 169ff.

Feasible network designs. We assume that a set $\mathcal{I}$ of fully parameterized potential antenna configurations is given. Each configuration is described by a complete set of parameters as listed in Def. 1 on p. 26, notably channel gain information. A selection of antenna configurations is represented by an incidence vector $z \in \{0,1\}^{\mathcal{I}}$, which we call a *network design* or *network configuration*. The value of the i-th component of z indicates whether or not configuration i is used in the network design ($z_i = 1$ or $z_i = 0$).

We furthermore assume that a set $\mathcal{F} \subset \{0,1\}^{\mathcal{I}}$ of *feasible network designs* is specified that contains the admissible combinations of configurations. In the capacity maximization model developed below, each antenna configuration is assigned to one sector, and $\mathcal{F}$ contains the requirement that exactly one configuration per sector be adopted; this is formalized in constraint (F) in Listing 1 on p. 121.

5.3.1 Simple objectives: cost and coverage

Network cost. In our simple model of network cost, a price $v_i \geq 0$ is associated to each configuration option $i \in \mathcal{I}$. The cost occurs if the configuration is selected. The following model describes the cost minimization problem:

$$\begin{aligned} \min \quad & \textstyle\sum_{i\in\mathcal{I}} v_i z_i \\ \text{s.t.} \quad & z \in \mathcal{F} \end{aligned} \tag{5.1}$$

An optimal solution to the problem (5.1) is $z = \mathbf{0}$, if $\mathbf{0} \in \mathcal{F}$—it is, after all, cheapest to not install any radio network. A more complex model may be needed to express a given cost structure in practice. For example, the cost of equipping a site with 3G hardware occurs only once, even if several sectors are installed there. Cases like this can require additional variables in the optimization formulation, but they are usually not difficult to model; we omit this for the sake of clarity.

Modeling coverage. The benign structure of traditional coverage problems is complicated by the notion of E_c/I_0 coverage. Most pre-UMTS coverage definitions are equivalent to E_c coverage. The underlying set-covering structure is complex in theory, but good, approximate solutions can be found with simple procedures; integer programming software tools are continuously improved in this regard. The notion of E_c/I_0 coverage, however, ties coverage to load. Optimization models including E_c/I_0 coverage are therefore inherently more involved.

We disentangle coverage and capacity and optimize only E_c coverage for the sake of model simplicity. We use only the deterministic path loss and disregard shadowing. The coverage of a network design can thus be determined directly from the input data and the pilot power level, without having to consider traffic, transmit powers, and stochastics. Even though not directly addressed, E_c/I_0 coverage is indirectly targeted through interference reduction in our capacity optimization scheme; this improves coverage in the cases studies discussed below.

Subdivisions of the planning area into pixels. Network coverage is commonly determined by checking the coverage condition on a discrete representation of the planning area through *pixels*. These are typically squares with a uniform side length in the range of 5–100 m; on each pixel, a representative channel attenuation value for each configuration is specified. Standard coverage models check the E_c-coverage condition (3.49) for each pixel. For obtaining a more efficient optimization models, pixels should be aggregated wherever possible. Formally, this means that we should strive to find a minimal subdivision of the area.

Figure 5.5: Minimal subdivisions of the planning area—example with three configuration options

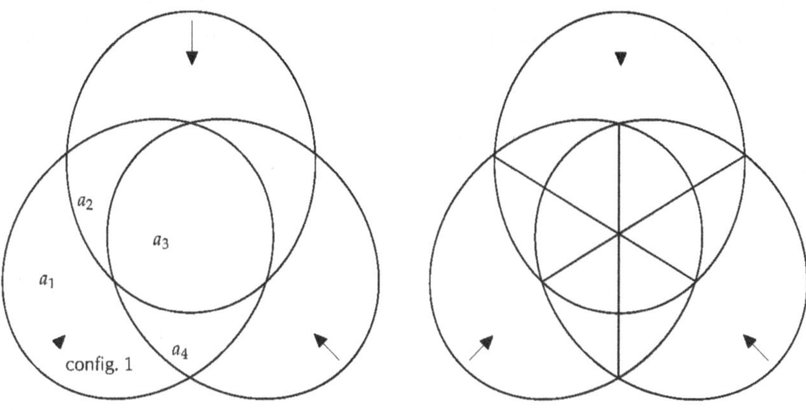

(a) Subdivision for the coverage model

(b) Finer subdivision for the capacity model

Let us formalize the condition for a subdivision of the planning area A to be suitable for determining coverage. A point in A can be covered by a configured sector $i \in \mathcal{I}$, if it lies in the *coverage set*

$$C_i := \{x \in A \mid \gamma_i^{\downarrow}(x) p_i^{(\mathrm{P})} \geq \pi_{\mathrm{r_c}}\}. \tag{5.2}$$

For a network design z, the total covered area is the union of the coverage set of all selected configurations:

$$A^{(\mathrm{r_c})}(z) := \bigcup_{\{i \,:\, z_i = 1\}} C_i. \tag{5.3}$$

A subdivision $\mathcal{P} \subset 2^A$ serves for determining the covered area for any network design $z \in 2^{\mathcal{I}}$, if the union of the elements covers the entire area, and, in addition, any $a \in \mathcal{P}$ is either contained in or disjoint from any set of the above form:

$$A^{(\mathrm{r_c})}(z) \cap a \in \{\emptyset, a\} \qquad \text{for all } a \in \mathcal{P},\ z \in 2^{\mathcal{I}}. \tag{5.4}$$

If the condition holds, then it suffices to check coverage for each element of $\mathcal{P}$ to determine $A^{(\mathrm{r_c})}(z)$.

The set of pixels in the representation of the propagation data meets the condition for suitable subdivisions, but there may be smaller ones. The subdivision with the least number of elements possible is the set of atoms of the set algebra generated by $A \cap \{C_i\}_{i \in \mathcal{I}}$ (*Sazonov*, 1988). We use this subdivision and denote it by $\mathcal{P}$; we continue to call its elements *pixels*. The minimal subdivision for an example with three configurations is illustrated in Fig. 5.5(a). The arrows indicate antennas; each antenna has an oval coverage area; each enclosed area corresponds to a pixel. Note that, unlike in the figure, a pixel in our sense needs not be connected.

Formal coverage model. We introduce a *coverage variable* $y_a \in \{0,1\}$ for each pixel $a \in \mathcal{P}$ to decide whether or not it is covered by any of the selected configurations ($y_a = 1$ or $y_a = 0$). This semantic is expressed by translating (5.3) into boolean arithmetics:

$$y_a = \bigvee_{i\,:\,a \subset C_i} z_i\,. \tag{5.5}$$

The operator $\vee$ stands for the logical disjunction "or".

The complete optimization model for coverage is derived by summing up the coverage contributions from the individual pixels with a suitable coefficient $f_a^{(c)} \geq 0$ in the objective. Furthermore, (5.5) is represented by a linear constraint:

$$\begin{aligned} \max \quad & \textstyle\sum_{a \in \mathcal{P}} f_a^{(c)} y_a && && (5.6a)\\ \text{s.t.} \quad & \textstyle\sum_{C_i \supset a} z_i \geq y_a && \text{for all } a \in \mathcal{P} && (5.6b)\\ & y \in \{0,1\}^{\mathcal{P}} \\ & z \in \mathcal{F} \end{aligned}$$

The precise notion of coverage depends on the choice of the coefficients $f_a^{(c)}$. The objective (5.6a) represents the covered area $|A^{(E_c)}|$ for $f_a^{(c)} := |a|$ and the covered traffic load $|A^{(E_c)}|_\ell^\downarrow$ for $f_a^{(c)} := \int_a T_\ell^\downarrow(x)\mathrm{d}x$. In the following, we use the shorthand notation

$$f^{(c)}(z) := \textstyle\sum_{a \in \mathcal{P}} f_a^{(c)} y_a \tag{5.7}$$

for the coverage level attained for a given network design z and calculated by a model of the above type.

Additional relations can be exploited to strengthen the formulation (5.6). Fig. 5.5(a) shows that if, for example, a_1 is covered, then configuration 1 must have been selected, so necessarily the pixels a_2, a_3, and a_4 are covered, too. We can hence add the inequalities

$$y_{a_2} \geq y_{a_1}, \quad y_{a_3} \geq y_{a_1}, \quad y_{a_4} \geq y_{a_1}$$

to the model. In a similar fashion, if pixel a_2 is covered, then a_3 must be covered, too. The explicit addition of inequalities of the type

$$y_a \leq y_b \quad \text{for all } a, b : \{i \mid a \subset C_i\} \subset \{i \mid b \subset C_i\}$$

considerably strengthens the MIP formulation of the model (5.6).

5.3.2 Capacity optimization

Three performance indicators reflect network capacity: grade of service, other-to-own-cell interference ratio, and cell load. We propose to focus on optimizing cell load.

Let us briefly discuss why we discarded the other capacity-related performance indicators. The grade of service is arguably the most important performance indicator, because it is linked directly to the user experience. If all users can be served (almost) all the time, then the operator needs not worry. There are, however, some arguments against using it as the primary objective for capacity: First, an informative value is difficult to estimate with a simple deterministic model, as we have seen in Sec. 4.4.2. Second, the grade of service is a bottleneck measure and can only be used sensibly in heavily congested cells. Reducing the other-to-own-cell interference ratio improves capacity; for example, *Sobczyk* (2005) argues that this improves the grade of service. It is, however, misguiding if used without relating it to the traffic served by the cell. After all, heavily loaded cells automatically have a limited other-to-own-cell interference ratio value; this is expressed in (3.39) and indicated in Fig. 3.4.

We minimize transmit power as an equivalent to cell load. For a given coupling matrix $C^{\downarrow}$, the transmit powers can be calculated with the Neumann series introduced in Sec. 3.1 as

$$\bar{p}^{\downarrow} = \textstyle\sum_{k=0}^{\infty} \left(C^{\downarrow}\right)^k \left(p^{(c)} + p^{(\eta)}\right)$$

as long as no cell is in overload. If overload occurs, then the transmit power is clipped at $\bar{p}^{\downarrow}_{\mathrm{max}}$. For capacity maximization, however, we abolish the clipping to put emphasis on avoiding congested cells. Overload situations disproportionately increase the components of the vector calculated in this way, so avoiding overload receives a high priority. Furthermore, we use the expected interference coupling matrix (assuming the restricted random model excluding shadowing, cf. Def. 4 on p. 30 and Lemma 4.1). In this chapter, the symbol $C^{\downarrow}$ thus denotes the expected coupling matrix throughout. This leads to the following optimization model for capacity:

$$\begin{aligned}
\min \quad & \textstyle\sum_{k=0}^{\infty} \left(C^{\downarrow}(z)\right)^k \left(p^{(c)} + p^{(\eta)}(z)\right) && (5.8a)\\
\text{s.t.} \quad & C^{\downarrow}(z) \text{ is the expected coupling matrix for } z && (5.8b)\\
& p^{(\eta)}(z) \text{ is the expected noise load for } z && (5.8c)\\
& z \in \mathcal{F}
\end{aligned}$$

We will provide formal representations of the constraints (5.8b) and (5.8c) below. Furthermore, the model is formulated with a vectorial objective function, for which we discuss a scalarization below.

In experiments, the optimization model (5.8) has turned out effective, because it combines load balancing and interference reduction aspects. It furthermore is a global measure that imparts high priority to overloaded areas and yet rewards interference reduction in mildly loaded areas. This makes it easier for search methods to improve configurations of adjacent cells with high load discrepancy.

5.3.3 Mixed integer linear programming model for capacity

We derive a novel integer programming model for capacity maximization; it is based on a description of the interference coupling matrix that is linear in the network design variables z. For the nonlinear objective (5.8a), we specify a tractable approximation.

Cell areas and subdivision for capacity optimization. To calculate the expected coupling matrix according to Lemma 4.1, we need to determine the cell areas A_i defined in (4.1). This requires a finer subdivision of the area than for coverage (cf. p. 112). We define the *dominance set* of i over j to be the set in which i provides a stronger average pilot signal than j:

$$\mathcal{D}_{ij} := \{x \in \mathcal{C}_i \mid p_i^{(\mathrm{P})}\gamma_i^{\downarrow}(x) > p_j^{(\mathrm{P})}\gamma_j^{\downarrow}(x)\}\,. \tag{5.9}$$

For a given network design z with $z_i = 1$, we have

$$A_i(z) = \mathcal{C}_i \setminus \left(\bigcup_{\{j\,:\,z_j=1\}} \mathcal{D}_{ji}\right) = \mathcal{C}_i \cap \overline{\bigcup_{\{j\,:\,z_j=1\}} \mathcal{D}_{ji}} = \mathcal{C}_i \cap \bigcap_{\{j\,:\,z_j=1\}} \overline{\mathcal{D}_{ji}}\,, \tag{5.10}$$

where for any set S, $\overline{S}$ denotes the complement $A \setminus S$. Any subdivision $\mathcal{P}$ for capacity optimization has to be compliant with all possible sets $A_i(z)$ in the sense analogous to condition (5.4). This is the case for the atoms of the smallest set algebra containing the families of sets $\{\mathcal{C}_i\}_{i\in\mathcal{I}}$ and $\{\mathcal{D}_{ij}\}_{i,j\in\mathcal{I}}$; Fig. 5.5(b) illustrates the concept.

We introduce an additional *service variable* $u_{ia} \in \{0,1\}$ expressing whether or not a pixel a is served by a given configuration i with $a \subset \mathcal{C}_i$. If the best server assumption holds, we can calculate the value of u_{ia} for any network design z by using (5.10):

$$u_{ia} = z_i \wedge \bigwedge_{\{j\in\mathcal{I}\,:\,a\subset\mathcal{D}_{ji}\}} \overline{z_j}\,. \tag{5.11}$$

Here, $\overline{z_j}$ denotes the boolean negation of z_j, *i.e.*, $(1 - z_j)$, and "$\wedge$" denotes the logical conjunction "and". The canonical linear formulation of (5.11) is:

$$\begin{aligned} u_{ia} &\le z_i \\ u_{ia} &\le 1 - z_j && \text{for all } j \in \mathcal{I} : a \subset \mathcal{D}_{ji} \\ u_{ia} &\ge z_i - \textstyle\sum_{\{j\in\mathcal{I}:a\subset\mathcal{D}_{ji}\}} z_j \end{aligned}$$

Additional inequalities strengthen the formulation analogous to the coverage model. The area $A_i(z)$ of cell i in a network design z is the set

$$A_i(z) = \bigcup_{a:u_{ia}=1} a\,.$$

Determining the entries of the interference-coupling matrix. We determine the entries of the expected coupling matrix according to Lemma 4.1; we use the notation introduced in Def. 4 on p. 30 and Sec. 4.2.1. For the main diagonal entries and the noise load vector, we add up the service variables with appropriate coefficients:

$$\begin{aligned} c_{ii}^{\downarrow}(z) &= \textstyle\sum_{a\subset A_i}\left(\int_a \omega(x) T_\ell^{\downarrow}(x)\mathrm{d}x\right) u_{ia}\,, \\ p_i^{(\eta)}(z) &= \textstyle\sum_{a\subset A_i}\left(\int_a \frac{\eta^{\downarrow}(x)}{\gamma_i^{\downarrow}(x)} T_\ell^{\downarrow}(x)\mathrm{d}x\right) u_{ia}\,. \end{aligned}$$

For calculating the off-diagonal elements, we introduce additional *interference variables* $v_{ia}^{(j)} \in \{0,1\}$. For any $a \in \mathcal{P}$ and $i, j \in \mathcal{I}$, they describe whether or not interference for cell i is generated by cell j at a. Interference is generated if the pixel is served by i and configuration j is selected:

$$v_{ia}^{(j)} = u_{ia} \wedge z_j\,.$$

This constraint is linearized as

$$v_{ia}^{(j)} \leq u_{ia}\,,$$
$$v_{ia}^{(j)} \leq z_j\,,$$
$$v_{ia}^{(j)} \geq u_{ia} + z_j - 1\,.$$

Because we minimize interference, the first two constraints are dispensable. The off-diagonal elements of the coupling matrix are calculated as weighted sums of the interference variables:

$$c_{ij}^{\downarrow}(z) = \sum_{a \subset A_i} \Big(\int_a \frac{\gamma_j^{\downarrow}(a)}{\gamma_i^{\downarrow}(a)} T_{\check{\ell}}(x) \mathrm{d}x \Big)\, v_{ia}^{(j)} \tag{5.13}$$

The complete set of constraints for the placeholders (5.8b) and (5.8c) is repeated as constraints (B) and (D) in Listing 1.

Linearization of interference coupling. Chances for solving large optimization problems are best for linear formulations, so we approximate the nonlinear objective (5.8) with a linear function. From the quantities that feature in the objective, the common channel power $p^{(c)}$ is a parameter; the coupling matrix $C^{\downarrow}$ and the noise load vector $p^{(\eta)}$ depend on the network design z. We use the first degree Taylor expansion (*Heuser*, 1995, Ch. 168) of the objective for $C^{\downarrow}$ and $p^{(\eta)}$ together with their linear description developed above to linearize the objective. The necessary partial derivatives can be calculated as follows (the symbol e_k denotes the k-th unit vector below):

Lemma 5.1: *Define for any $C^{\uparrow} \in \mathbb{R}_+^{n \times n}$ and $p^{(c)}, p^{(\eta)} \in \mathbb{R}_+^n$ the function*

$$\bar{p}^{\downarrow}\Big(C^{\downarrow}, p^{(c)}, p^{(\eta)}\Big) := \sum_{k=0}^{\infty} (C^{\downarrow})^k \Big(p^{(c)} + p^{(\eta)}\Big)\,.$$

Then $\bar{p}^{\downarrow}$ is differentiable wherever $\rho(C^{\uparrow}) < 1$ and the partial derivatives are

$$\frac{\partial \bar{p}^{\downarrow}}{\partial c_{k\ell}^{\downarrow}} = \Big(\frac{\partial \bar{p}_i^{\downarrow}}{\partial c_{k\ell}^{\downarrow}}\Big)_{1 \leq i \leq n} = \big(I - C^{\downarrow}\big)^{-1} \cdot e_k \cdot \bar{p}_\ell^{\downarrow} \qquad \textit{for all } 1 \leq k, \ell \leq n$$

and

$$D_{p^{(\eta)}} \bar{p}^{\downarrow} = \Big(\frac{\partial \bar{p}_i^{\downarrow}}{\partial p_k^{(\eta)}}\Big)_{1 \leq i,k \leq n} = \big(I - C^{\downarrow}\big)^{-1}\,.$$

Proof. The derivatives can be calculated based on the coupling equation system (3.13)

$$\bar{p}_i^{\downarrow} = \sum_j c_{ij}^{\downarrow} \bar{p}_j^{\downarrow} + p_i^{(\eta)} + p_i^{(c)}\,.$$

Differentiating this for a given entry $c_{k\ell}^{\downarrow}$ of the coupling matrix, we obtain:

$$\frac{\partial \bar{p}_i^{\downarrow}}{\partial c_{k\ell}^{\downarrow}} = \sum_j \frac{\partial (c_{ij}^{\downarrow} \bar{p}_j^{\downarrow})}{\partial c_{k\ell}^{\downarrow}} = \sum_j c_{ij}^{\downarrow} \frac{\partial \bar{p}_j^{\downarrow}}{\partial c_{k\ell}^{\downarrow}} + \delta_{ik} \bar{p}_l^{\downarrow}\,.$$

Here, δ_{ik} denotes the Kronecker symbol ($\delta_{ik} = 1$ if $i = k$, $\delta_{ik} = 0$ otherwise). Similar to the power values, their partial derivatives are governed by the equation system

$$\frac{\partial \bar{p}^{\downarrow}}{\partial c^{\downarrow}_{k\ell}} = C^{\downarrow} \cdot \frac{\partial \bar{p}^{\downarrow}}{\partial c^{\downarrow}_{k\ell}} + e_k \cdot \bar{p}^{\downarrow}_{\ell}$$

and can be explicitly calculated as

$$\frac{\partial \bar{p}^{\downarrow}}{\partial c^{\downarrow}_{k\ell}} = \left(I - C^{\downarrow}\right)^{-1} \cdot e_k \cdot \bar{p}^{\downarrow}_{\ell}\,.$$

When differentiating for the noise load $p^{(\eta)}$, we obtain in a similar fashion

$$\frac{\partial \bar{p}^{\downarrow}_i}{\partial p^{(\eta)}_k} = \sum_j \frac{\partial (c^{\downarrow}_{ij} \bar{p}^{\downarrow}_j)}{\partial p^{(\eta)}_k} + \delta_{ik} = \sum_j c^{\downarrow}_{ij} \frac{\partial \bar{p}^{\downarrow}_j}{\partial p^{(\eta)}_k} + \delta_{ik}\,,$$

and thus

$$\frac{\partial \bar{p}^{\downarrow}}{\partial p^{(\eta)}_k} = \left(I - C^{\downarrow}\right)^{-1} \cdot e_k\,,$$

which proves the lemma. □

For a given initial network state specified by $C^{(0)} = \left(c^{(0)}_{ij}\right)_{i,j} \in \mathbb{R}^{n\times n}_{+}$ and $p^{(\eta,0)} \geq 0$, we use the shorthand notation

$$\bar{p}^{(0)} := \left(I - C^{(0)}\right)^{-1} \left(p^{(C)} + p^{(\eta,0)}\right)$$

and obtain a linear approximation $\tilde{p}^{\downarrow}$ of $\bar{p}^{\downarrow}$ at $C^{(0)}$ and $p^{(\eta,0)}$ through the Taylor expansion for vectorial functions (*Heuser*, 1995, Th. 168.4) as

$$\begin{aligned}
\tilde{p}^{\downarrow}\left(C^{\downarrow}, p^{(\eta)}\right) &= \bar{p}^{(0)} + \sum_{k,\ell} \frac{\partial \bar{p}^{(0)}}{\partial c^{\downarrow}_{k\ell}} \left(c^{\downarrow}_{k\ell} - c^{(0)}_{k\ell}\right) + D_{p^{(\eta)}} \bar{p}^{(0)} \left(p^{(\eta)} - p^{(\eta,0)}\right) \\
&= \bar{p}^{(0)} + \left(I - C^{(0)}\right)^{-1} \left(C^{\downarrow} - C^{(0)}\right) \bar{p}^{(0)} + \left(I - C^{(0)}\right)^{-1} \left(p^{(\eta)} - p^{(\eta,0)}\right).
\end{aligned}$$

For using this as an objective in optimization, we may remove the constant parts and use the shorter version

$$\left(I - C^{(0)}\right)^{-1} \cdot \left(C^{\downarrow} \bar{p}^{(0)} + p^{(\eta)}\right).$$

Even for a limited set of configuration options, the cell structure and the coupling matrix of a network can change drastically. A linear approximation that is based on the coupling matrix of an initial network configuration has therefore turned out to be bad for most other configurations. For our model aiming at global capacity optimization, we therefore choose the linearization for $c^{(0)}_{ij} = 0$ and $p^{(\eta,0)} = 0$. This amounts to:

$$\begin{aligned}
\min \quad & C^{\downarrow}(z) p^{(C)} + p^{(\eta)}(z) && \text{(5.14a)} \\
\text{s.t.} \quad & C^{\downarrow}(z) \text{ is the expected coupling matrix for } z && \text{(5.14b)} \\
& p^{(\eta)}(z) \text{ is the expected noise load for } z && \text{(5.14c)} \\
& z \in \mathcal{F}
\end{aligned}$$

This linearization led to best results in experiments with on our specific datasets, but the best approximation point may depend on the structure of $\mathcal{F}$.

Convex approximation of interference coupling. Also convex functions are manageable in large-scale optimization. If we neglect noise, Lemma 3.9 suggests a sensible convex approximation of the capacity objective. If we substitute the other-to-own-cell interference ratio $\bar{\iota}_i^{\downarrow}$ by a fixed estimate $\hat{\iota}^{\downarrow}$ in the generalized pole equation, the i-th component of the objective function becomes a convex function of the main diagonal element $c_{ii}^{\downarrow}$, while the off-diagonal elements are eliminated:

$$\begin{aligned} \min \quad & \left(\tfrac{p_i^{(c)}}{1-(1+\hat{\iota}^{\downarrow})c_{ii}^{\downarrow}}\right)_{i\in\mathcal{I}} && (5.15a) \\ \text{s.t.} \quad & C^{\downarrow}(z) \text{ is the expected coupling matrix for } z && (5.15b) \\ & z \in \mathcal{F} \end{aligned}$$

For using linear integer programming solvers, we approximate (5.15a) by a piecewise linear convex function. In the complete listing on p. 121, this approximation is realized through additional variables $d_i \geq 0$ and the constraints (C) for each linear segment. The parameters are chosen such that for each $i \in \mathcal{I}$ the segments represent the tangent to the graph of the function

$$c_{ii}^{\downarrow} \mapsto p_i^{(c)}/1 - (1+\hat{\iota}^{\downarrow})c_{ii}^{\downarrow}.$$

The linear and the convex approximation reflect different aspects of interference reduction. The linear model targets inter-cell interference, whereas the convex approximation strives to evenly distribute load among cells and reduce intra-cell interference. It has turned out that both are best used in conjunction; the data discussed in Sec. 5.5.2 present some evidence. We use a weighted sum of (5.14a) and (5.15b) as the objective for the convex integer programming model. The weight parameter is denoted by $\beta > 0$, and we obtain:

$$\begin{aligned} \min \quad & \beta\big(C^{\downarrow}(z)p^{(c)} + p^{(\eta)}(z)\big) + (1-\beta)\left(\tfrac{p_i^{(c)}}{1-(1+\hat{\iota}^{\downarrow})c_{ii}^{\downarrow}}\right)_{i\in\mathcal{I}} && (5.16a) \\ \text{s.t.} \quad & C^{\downarrow}(z) \text{ is the expected coupling matrix for } z && (5.16b) \\ & p^{(\eta)}(z) \text{ is the expected noise load for } z && (5.16c) \\ & z \in \mathcal{F} \end{aligned}$$

Because the linear approximation is always an underestimation, values of $\beta > 0.5$ have turned out best in experiments; in the case studies below we use $\beta = 0.8$.

5.3.4 Combined optimization models

We obtain three optimization problems by applying the ϵ-constraint scalarization method and focusing on cost, coverage, or capacity. For the sake of clarity, the notation is held at symbolic level here; the concrete interpretation of the combined capacity optimization model (5.19) used in the case studies is detailed in Listing 1 on p. 121. The actual results achieved with these models and their tractability depend largely on the chosen structure of the feasible set $\mathcal{F}$ and the constraints on the secondary objectives. We will specify our choices in Sec. 5.4 below.

Maximize coverage. The limit value for network cost is denoted by $N_{\max} \geq 0$. Capacity is constrained per cell with the vector $\boldsymbol{\psi} \geq \mathbf{0}$. The optimization model for coverage maximization is thus:

$$\begin{aligned} \max \quad & f^{(c)}(\boldsymbol{z}) && (5.17a) \\ \text{s.t.} \quad & \textstyle\sum_{i \in \mathcal{I}} \nu_i z_i \leq N_{\max} && (5.17b) \\ & \textstyle\sum_{k=0}^{\infty} \left(\boldsymbol{C}^{\downarrow}(\boldsymbol{z})\right)^k \left(\boldsymbol{p}^{(c)} + \boldsymbol{p}^{(\eta)}(\boldsymbol{z})\right) \leq \boldsymbol{\psi} && (5.17c) \\ & \boldsymbol{z} \in \mathcal{F} \end{aligned}$$

Expression (5.17a) represents the coverage objective, constraint (5.17b) limits network cost, and constraints (5.17c) limit individual cell load.

Minimize cost. We specify the constraint on coverage in relative terms through a fraction $\zeta_{\min}^{(E_c)} \in [0,1]$ of a reference coverage value $K^{(cov)} > 0$. The optimization model for cost minimization is:

$$\begin{aligned} \min \quad & \textstyle\sum_{i \in \mathcal{I}} \nu_i z_i && (5.18a) \\ \text{s.t.} \quad & f^{(c)}(\boldsymbol{z}) \geq \zeta_{\min}^{(E_c)} K^{(cov)} && (5.18b) \\ & \textstyle\sum_{k=0}^{\infty} \left(\boldsymbol{C}^{\downarrow}(\boldsymbol{z})\right)^k \left(\boldsymbol{p}^{(c)} + \boldsymbol{p}^{(\eta)}(\boldsymbol{z})\right) \leq \boldsymbol{\psi} && (5.18c) \\ & \boldsymbol{z} \in \mathcal{F} \end{aligned}$$

Minimize interference. We propose to use unit weights to scalarize the multi-objective optimization model (5.8) for obtaining a single-objective capacity maximization model:

$$\begin{aligned} \min \quad & \mathbf{1}^t \textstyle\sum_{k=0}^{\infty} \left(\boldsymbol{C}^{\downarrow}(\boldsymbol{z})\right)^k \left(\boldsymbol{p}^{(c)} + \boldsymbol{p}^{(\eta)}(\boldsymbol{z})\right) && (5.19a) \\ \text{s.t.} \quad & f^{(c)}(\boldsymbol{z}) \geq \zeta_{\min}^{(E_c)} K^{(cov)} && (5.19b) \\ & \textstyle\sum_{i \in \mathcal{I}} \nu_i z_i \leq N_{\max} && (5.19c) \\ & \boldsymbol{z} \in \mathcal{F} \end{aligned}$$

Listing 1 contains the complete mixed integer linear programming model for interference minimization that is used in the computational case studies. The objective (A) is the convex approximation (5.16a) of the objective, scalarized with unit weights. Constraint (5.19b) corresponds to the inequality (E) in the listing. Constraint (5.19c) is omitted in Listing 1, because in the experiments no additional cost for adjusting an antenna is assumed; the number of cells is fixed by constraint (F). The constraint $\boldsymbol{z} \in \mathcal{F}$ is interpreted here as the choice for one of the potential configurations per cell being mandatory; it is formalized in constraint (F).

5.4 Computational case studies

We now demonstrate how the above optimization models can be used in realistic settings to effectively improve network performance and notably maximize network

Listing 1 Complete MIP for capacity optimization under coverage constraint

$$\min \quad \beta \sum_{i,j \in \mathcal{I}} p_i^{(c)} c_{ij}^{\downarrow} + \beta \sum_{i \in \mathcal{I}} p_i^{(\eta)} + (1-\beta) \sum_{i \in \mathcal{I}} d_i \qquad \text{(A)}$$

$$\left.\begin{aligned}
\text{s.t.} \quad & c_{ii}^{\downarrow} = \sum_{a \subset A_i} \left(\textstyle\int_a \omega(x) T_\ell^{\downarrow}(x) \mathrm{d}x \right) u_{ia} && \text{for all } i \in \mathcal{I} \\
& c_{ij}^{\downarrow} = \sum_{a \subset A_i} \left(\textstyle\int_a \frac{\gamma_j^{\downarrow}(x)}{\gamma_i^{\downarrow}(x)} T_\ell^{\downarrow}(x) \mathrm{d}x \right) v_{ia}^{(j)} && \text{for all } i,j \in \mathcal{I},\ i \neq j \\
& p_i^{(\eta)} = \sum_{a \subset A_i} \left(\textstyle\int_a \frac{\eta^{\downarrow}(x)}{\gamma_i^{\downarrow}(x)} T_\ell^{\downarrow}(x) \mathrm{d}x \right) u_{ia} && \text{for all } i \in \mathcal{I}
\end{aligned}\right\} \text{(B)}$$

$$d_i \geq \delta_{i,t}^{(o)} + \delta_{i,t}^{(s)} c_{ii}^{\downarrow} \qquad \text{for all } t \in \mathcal{G}, i \in \mathcal{I} \qquad \text{(C)}$$

$$\left.\begin{aligned}
& u_{ia} \leq z_i && \text{for all } i \in \mathcal{I}, a \in \mathcal{P},\ C_i \supset a \\
& u_{ia} \leq 1 - z_j && \text{for all } i,j \in \mathcal{I}, a \in \mathcal{P},\ a \subset \mathcal{D}_{ji} \\
& u_{ia} \geq z_i - \sum_{\{j \in \mathcal{I}: a \subset \mathcal{D}_{ji}\}} z_j && \text{for all } i \in \mathcal{I}, a \in \mathcal{P},\ C_i \supset a \\
& v_{ia}^{(j)} \geq u_{ia} + z_j - 1 && \text{for all } i,j \in \mathcal{I},\ a \in \mathcal{P},\ C_i \supset a
\end{aligned}\right\} \text{(D)}$$

$$\sum_{i \in \mathcal{I}} \sum_{a \subset A_i} \left(\textstyle\int_a T_\ell^{\downarrow}(x) \mathrm{d}x \right) u_{ia} \geq \zeta_{\min}^{(E_c)} K^{(\mathrm{cov})} \qquad \text{(E)}$$

$$\sum_{i \in \mathcal{I}_s} z_i = 1 \qquad \text{for all } s \in \mathcal{S} \qquad \text{(F)}$$

$$\begin{aligned}
& u_{ia} \in \{0,1\} && \text{for all } i \in \mathcal{I}, a \in \mathcal{P},\ C_i \supset a \\
& v_{ia}^{(j)} \in \{0,1\} && \text{for all } i,j \in \mathcal{I},\ a \in \mathcal{P},\ C_i \supset a \\
& z_i \in \{0,1\} && \text{for all } i \in \mathcal{I} \\
& c_{ij}^{\downarrow} \geq 0 && \text{for all } i,j \in \mathcal{I} \\
& p_i^{(\eta)} \geq 0 && \text{for all } i \in \mathcal{I} \\
& d_i \geq 0 && \text{for all } i \in \mathcal{I}
\end{aligned}$$

Annotations and additional notation

(C) Variables d_i approximate the convex objective components. Each inequality is a tangent in a point from $\mathcal{G} := \{i/11(1+\hat{\imath}^{\downarrow}), i = 0, \dots, 10\}$ with slope and ordinate

$$\delta_{i,t}^{(s)} := \frac{p_i^{(c)}(1+\hat{\imath}^{\downarrow})}{(1-(1+\hat{\imath}^{\downarrow})t)^2}, \qquad \delta_{i,t}^{(o)} := -t\,\delta_{i,t}^{(s)} + \frac{p_i^{(c)}}{1-(1+\hat{\imath}^{\downarrow})t} \qquad t \in \mathcal{G}, i \in \mathcal{I}.$$

(E) Coverage constraint for a coverage reference value of $K^{(\mathrm{cov})} \geq 0$.

(F) Description of feasible combinations. For a set $\mathcal{S}$ of cells, the set $\mathcal{I}$ is partitioned into $\{\mathcal{I}_s\}_{s \in \mathcal{S}}$. For each cell, exactly one configuration has to be chosen.

Table 5.2: Key properties of data scenarios used for case studies

	area [km²]	no. of sectors	resolu-tion [m][a]	Traffic [km⁻²]	traffic mix [%] voice	video	data	add. loss [dB][d]	vert. ang.[e]
Berlin	56.25	204	50	0.92	18.7	14.3	67.0	11	6.5°
Lisbon	21.00	164	20	2.50	18.2	14.4	67.4	11	6.0°
Turin	274.00	335[b]	50	0.35	48.7	44.4	6.9	8/20[c]	4.0°
Vienna	437.00	628[b]	50	0.22	48.4	0.0	51.6	15	15.0°

[a] Pixel length for grid representation of propagation data
[b] Sectors in urban area: 133 (Turin), 519 (Vienna)
[c] outdoor/indoor
[d] Maximum offset against base coverage threshold
[e] Antenna's vertical opening angle

capacity. Sec. 5.4.1 provides a brief overview of the four scenarios for which case studies are presented; we detail how we use the combined models and which solution methods we employ in Sec. 5.4.2. The results of optimization are discussed in Sec. 5.4.3. All computations have been made with the custom software components described in Ch. 1 and depicted in Fig. 1.2.

5.4.1 Realistic data scenarios

We use four scenarios based on realistic datasets of different sizes and characteristics. The data are similar to those used in practice. An overview of key information on all scenarios is presented in Tab. 5.2; App. A contains additional details. The Berlin and Lisbon datasets have been compiled by the consortium of the IST project MOMENTUM and are publicly available (*Eisenblätter et al.*, 2005; MOMENTUM *Project*, 2003). They feature medium-sized inner city settings. The Turin and Vienna cases encompass larger areas around the urban centers and have been compiled by the MORANS subworking group of the COST 273 initiative (*Munna et al.*, 2004; *Verdone & Buehler*, 2003). Both datasets have previously been used in scientific publications (e. g., *Eisenblätter et al.*, 2005a; *Lamers et al.*, 2003; *Siomina et al.*, 2006).

For each setting, an inhomogeneously distributed traffic intensity function composed of several services is given. The intensity function for Lisbon is depicted in Fig. 5.6(b); similar plots for the other cases are given in Figs. A.1(b), A.2(b), and A.3(b). The average traffic intensity is listed in Tab. 5.2; the values vary between 2.5 units per square kilometer in Lisbon and 0.22 km^{-2} in Vienna. Tab. 5.2 also provides a top-level view on the service mixes; the precise service models and parameters are listed in detail in Tabs. A.6 and A.5 in the appendix. The downlink is the bottleneck direction in most scenarios because downlink-biased data services have a large traffic share. Only in Turin, bidirectional services prevail, making a potentially uplink-limited scenario. We focus on the downlink in Turin nonetheless.

A number of candidate sites and sectors with a recommended initial configuration

Figure 5.6: Geographical data and traffic, Lisbon scenario

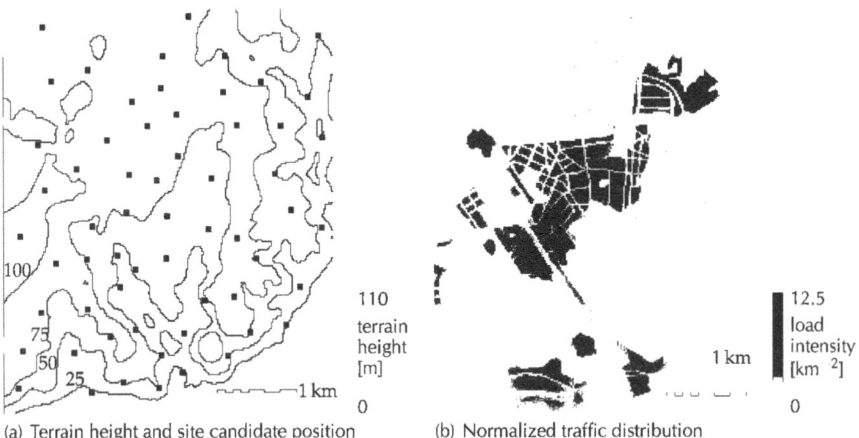

(a) Terrain height and site candidate position (b) Normalized traffic distribution

is available to serve the traffic. The candidate sites for Lisbon are shown in Fig. 5.6(a). The initial configuration specifies the maximum number of sectors to be used at a site and preferred tilt and azimuth values for each sector; uniform tilt values and regular sectorization are predominantly used. The initial configurations for Berlin and Lisbon are shown in Fig. 5.8. The total number of candidate sectors per scenario is listed in Tab. 5.2. We subdivide the sites in the MORANS scenarios into *rural* and *urban* ones. This distinction is depicted in Fig. 5.7. Rural sites will only be modified to improve coverage, whereas urban sites are also optimized for capacity.

For obtaining signal propagation information, real antenna diagrams with varying properties are superimposed onto (isotropic) COST-Hata path loss information. The path loss computation uses clutter-specific diffraction loss and diffraction angles based on terrain height information. Different clutter types are shown for Berlin in Fig. 2.7. The terrain height data for Lisbon is shown in Fig. 5.6(a); the same data for the other scenarios are contained in Figs. A.1(a), A.2(a), and A.3(a). The Berlin scenario is comparatively flat, whereas Lisbon exhibits large height differences over a small area. Both Turin and Vienna have their urban centers on an inclined plane and, in addition, feature mountainous areas. The vertical opening angles as shown in Tab. 5.2 vary with the antenna type. The type to be used per site is prescribed; there is one antenna type used per scenario except for Turin, where a small fraction of sectors deviates from the standard type. Tab. A.3 lists all antenna types and their parameters in detail. Note that we have used a smoothed version of the antenna diagrams for the MOMENTUM scenarios to avoid artefacts created by side lobes; the smoothed diagrams are shown in Fig. A.5.

Degrees of freedom and parameters for objectives. We consider site selection, antenna tilt, and azimuth as degrees of freedom. The total downtilt of a sector can be varied in

Figure 5.7: Urban areas (light shaded), optimization focus polygon, candidate sites (MORANS scenarios)

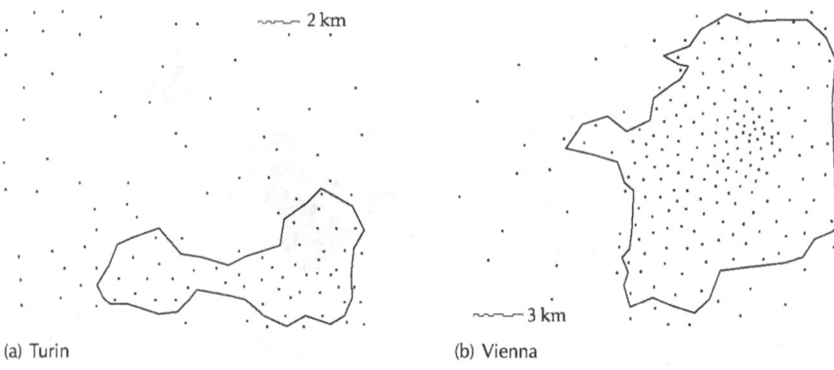

(a) Turin (b) Vienna

the range of 2–12° in steps of 2°.[4] A tilt of 0° potentially causes too much interference, but it is exceptionally admitted if used in a sector's initial configuration. The downtilt is realized as far as possible by electrical tilting to avoid side-lobe and back-lobe interference; mechanical tilt is only used for tilt values higher than the respective antenna's maximum electrical tilt value. The maximum electrical tilt values depend on the antenna type; they are specified in Tab. A.3. In addition to the tilt manipulation, a sector's azimuth may be varied by ±30° from the preferred direction. The pilot power is fixed at 33 dBm in Vienna and 30 dBm in all other scenarios. Each configuration option in $\mathcal{I}$ is assigned to a sector; we shall introduce notation for this relation: We denote the set of sectors by $\mathcal{S}$ and set of configuration options for a sector $s \in \mathcal{S}$ by $\mathcal{I}_s \subset \mathcal{I}$. We thus have $\mathcal{I} = \dot{\cup}_{s \subset \mathcal{S}} \mathcal{I}_s$.

The parameters determining the objective functions are as follows. In our cost model, each sector in use incurs uniform cost. The condition for E_c coverage is based on a common base threshold of −105 dBm with a scenario-specific offset. This offset varies for each user to reflect additional attenuation caused by vehicles or buildings or body loss. To ensure full coverage, we have imposed the maximum applicable offset listed in Tab. 5.2 in each scenario. Only in Turin, indoor and outdoor coverage are treated separately; details are given below. For E_c/I_0 coverage, a threshold of −15 dB is imposed. The downlink capacity limit is a maximum downlink average transmit power of $\bar{p}^{\downarrow}_{\max} = 14\,\mathrm{W} \simeq 11.5\,\mathrm{dBm}$.

5.4.2 Optimization method

Our strategy is to improve coverage, cost, and capacity in a hierarchical fashion in this order, in each step admitting different sets of configuration changes and employing a custom solution algorithm. The first two steps are geared to produce challenging

[4] In the Vienna scenario, the range is limited to 2–10°, because the "60 deg sector" antenna model does not feature electrical tilt and a maximum mechanical tilt of 10° is prescribed.

capacity optimization instances with comparatively high load; under realistic circumstances, a more conservative parameterization might be advisable to ensure sufficient capacity in the final configuration. The heuristic algorithms used for solving the involved optimization problems are detailed below. For the capacity optimization step, we compare the performance of local search and integer programming.

Hierarchical optimization regime. The capacity optimization step is the last of three steps. Because that step is our focus, the initial two steps of coverage maximization and cost minimization can here be seen as preprocessing.

(a) *Coverage maximization* is undertaken first by solving problem (5.17). All candidate installations are considered. The set $\mathcal{F}$ contains the constraint that per cell one installation be selected. In the inner-city scenarios Berlin and Lisbon, a uniform standard configuration for all sectors leads to full coverage, so the coverage maximization problem is trivial. Tilts of 2° are used only for coverage reasons and are therefore only admitted at this stage. We have not imposed hard constraints on cost and capacity ($\psi = \infty$, $N_{\mathsf{max}} = \infty$). The best coverage value achieved in this step is taken as the reference value $K^{(\mathrm{cov})}$ in the subsequent steps.

(b) Next, we *minimize cost.* We use only the numbers of sectors and sites as an indicator for network cost and try to minimize them while maintaining the level of coverage achieved in the first step. The problem is formally defined in (5.18). As coverage constraint, we impose the level $K^{(\mathrm{cov})}$ of coverage obtained in the first step relaxed by 0.5 % ($\zeta_{\min}^{(E_c)} = 0.995$). To reduce complexity, we do not admit any reconfigurations at this stage; the set $\mathcal{F}$ of admissible candidates contains only the configurations selected in the previous step, and either all installations from the same site have to be selected or none. Our optimization model (and the heuristic used to solve it described below) are comparatively simple; the previous literature discussed above contains more sophisticated approaches for this step.

   Our approach is not purely hierarchical, as the capacity limit ψ is already at this stage imposed on the interference function. This is necessary to prevent an excessive elimination of sites in the areas with high traffic intensity and little coverage issues. The constraint has been chosen as $\psi = 20\,\mathrm{W}$. This comparatively loose values produce highly loaded starting configurations for the last optimization step.

(c) Finally, *capacity* is maximized by minimizing the interference function as described in (5.19). For the sectors remaining after cost minimization, all configuration options except tilt values of 2° are admitted unless the low tilt value has been resorted to in coverage maximization. Again, $\mathcal{F}$ contains the constraint that exactly one configuration per cell be selected. The coverage limit is identical to the previous stage. In the Turin and Vienna scenarios, only the urban sites may be modified.

Greedy heuristic. The cost minimization problem (5.18) is solved by a greedy procedure. We first seek to identify and deactivate entire dispensable sites. All sites

Figure 5.8: Initial network configuration and results of preprocessing

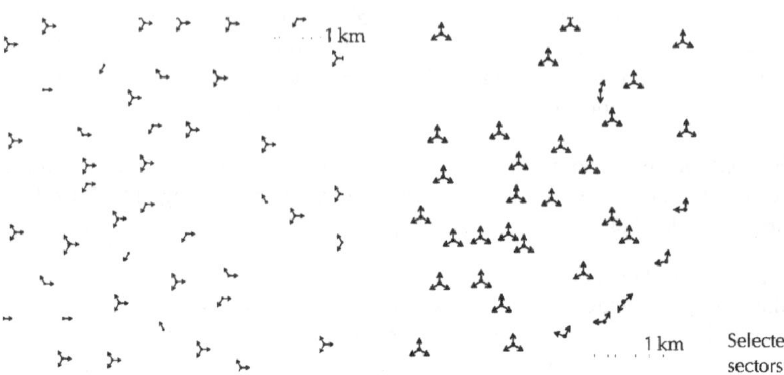

(a) Berlin (b) Lisbon

are sorted by ascending total load (in all associated cells) and sequentially tentatively removed from the network design. If the resulting coverage and load are in the permitted range and the reduced configuration is thus feasible for problem (5.18), then the site is permanently removed. Otherwise, the site remains active and is never reconsidered. In the Berlin scenario, a second step repeating this scheme but trying to deactivate individual sectors is used, starting from the cell with the lowest load. Because the first pass reduces overall interference, it may happen that all sectors of a site that survived the first round can be deactivated, thus disposing of the entire site even though it could not be deleted in the first round.

Local search. The coverage optimization (5.17) and capacity maximization (5.19) problems are solved heuristically by a local search procedure. In both cases, the algorithm is given a current network configuration and checks if some feasible move increases the objective function. If this is the case, then the move yielding most improvement in the objective is performed and the current configuration is adjusted accordingly. This is repeated until no improving feasible moves are found. The feasible moves are modifications of the configuration of a single antenna.

For achieving a competitive running time, it is essential to consider the moves in a structured fashion and to reduce the number of tentative network configurations evaluated. Even though the expected-coupling method can be implemented to run efficiently, a complete evaluation of all potential moves in each step would be too time consuming. To better focus the optimization efforts, a fitness score is calculated per cell. The score is influenced by coverage problems in the cell area, cell load, and the amount of blocked traffic. The score takes lower values for cells with performance problems. Optimization takes place in rounds. In each round, a threshold is fixed and cells with a score below this threshold are selected for optimization. The threshold is raised from round to round. This ensures that the most severe problems are treated first, whereas minor performance deficiencies have less priority. A sequence of cells used for the next optimization round is then obtained by sorting the selected cells according to their score and inserting the most influential neighbors of each selected cell directly after the cell itself into the sequence. The potential impact of a neighbor is estimated according to the corresponding coupling row. The alternative configurations are then tested for all cells in the sequence.

MIP heuristic. In the MIP heuristic, we solve the tractable approximation (5.16) of the interference function with integer programming methods. Because the complete simultaneous optimization of all sectors is impossible within reasonable time, we solve small subproblems derived by admitting modifications of a small number k of sectors at a time. The algorithm can thus be seen as a *k-opt heuristic* for capacity maximization; in all cases we set $k = 4$. Smaller subproblems have turned out to rarely produce improving configurations, whereas larger subproblems consume extensive computational time while producing only marginally better results. The advantage of the MIP heuristic over local search is that it is able to manipulate several sectors at a time while considering the effect of all surrounding cells (in the approximative sense).

A sequence of sets of four sectors to consider are derived as follows: For each cell that has more than average load, the three sectors with largest potential overlap are added. The sector sets are considered ordered by descending load. Potential overlap between two sectors $s, t \in \mathcal{S}$ is calculated as $\sum_{i \in \mathcal{I}_s, j \in \mathcal{I}_t} |C_i \cap C_j|^{\downarrow}_{\ell}$, *i. e.*, considering all potential configuration options for both cells. This is in general different from the neighbors causing most interference in the current coupling matrix. This scheme has turned out to be most effective in computational tests.

The resulting MIP is solved with the mathematical programming software CPLEX (v. 10.1, ILOG *S. A.*, 2006). During an optimization run, a handful of incumbent solutions are usually found. For each solution improving the approximative objective function (5.16a), the exact objective function (5.19a) is evaluated. Only if the exact function is also improved, the modifications are adopted. For the considered problems to be solvable in reasonable time, it is crucial to control their size and complexity. To this end, excluding low tilt values is sensible: They hardly occur in good solutions but have far-reaching consequences, which directly translate into model complexity. In addition to restricting the options, the size of the model is reduced in the following way: Interference variables $v_{ia}^{(j)}$ that have small contributions to the respective off-diagonal elements (*i. e.*, those variables with small coefficients in the right-hand side of (5.13)) and less important off-diagonal elements $c_{ij}^{\downarrow}(z)$ (*i. e.*, those elements for which the sum of *all* coefficients in the right-hand-side of (5.13) is small) are omitted.

The starting point to the MIP heuristic is the result of local search, because computational tests have shown that the selected subproblems then better focus on problematic areas. The solution of the MIP heuristic is thus never worse than the result of local search. The resulting configuration is in addition "polished" by an eventual additional call to the local search procedure.

Table 5.3: Result of preprocessing
(coverage maximization and site selection)

		E_c coverage [%]		infrastructure	
		area	traffic	no. of sites	no. of cells
Berlin	initial	100.0	100.0	68	204
	preprocessed	100.0	100.0	43	108
Lisbon	initial	100.0	100.0	60	164
	preprocessed	99.5	99.7	34	96
Turin	initial	81.5[a]	60.7	111	335
	preprocessed	86.2[a]	62.2	111	335
Vienna	initial	66.7	95.0	211	628
	preprocessed	68.9	95.3	122	361

[a] Outdoor coverage. A detailed account of coverage optimization considering also the stricter indoor constraint is given in Tab. 5.5.

5.4.3 Computational results

The main results of this section are the performance gains achieved by capacity optimization and the comparison of the two optimization methods used for this last optimization step; they are summarized in Tab. 5.4. An overview on the preprocessing steps providing the capacity optimization instance considered in each scenario is contained in Tab. 5.3. The results of preprocessing and capacity optimization are discussed below in detail for each scenario; we first explain the columns of the tables and the associated evaluation methods.

For each scenario, Tab. 5.3 lists the achieved coverage and the required network infrastructure before and after preprocessing. The first two columns contain the E_c coverage computed with medians of channel attenuation, *i.e.*, disregarding shadow fading (see Sec. 2.5.2 and Def. 2 on p. 26), in percentage of the area and percentage of the traffic load. The last two columns list the number of base station sites and the number of active cells. As most sites are three-sectorized, there are roughly three times as many cells as sites.

Tab. 5.4 contains the deterministic objective function of capacity maximization and a selection of essential downlink performance figures obtained through Monte Carlo simulation including random shadow fading. Some performance indicators have been modified such that smaller values are preferable throughout. The objective function value (5.19a) has the label "obj". The second column "$|A \setminus A^{(C)}|_\ell^\downarrow$" shows the expected uncovered traffic load. In contrast to Tab. 5.3, also E_c/I_0 coverage is checked and stochastic shadow fading is taken into account. Column "$\sum_i \bar{p}_i^\downarrow$" contains the total power emitted by the cells. It is equivalent to the expected average load "$\bar{L}^\downarrow$" in the next column multiplied with $n \cdot p_{\max}^\downarrow$ because we consider only macro cells with a uniform maximum power value $p_{\max}^\downarrow$. In addition, we specify the maximum expected cell load "$\max_i L_i^\downarrow$". Column "$\bar{\iota}^\downarrow$" contains the expected average other-to-own-cell in-

Table 5.4: Results of capacity optimization: means of downlink performance indicators[a]

		obj [W]	$\lvert A\setminus A^{(C)}\rvert_{\ell}^{\downarrow}$ [‰]	$\sum_i \bar{p}_i^{\downarrow}$ [W]	$\bar{L}^{\downarrow}$ [%]	$\max_i L_i^{\downarrow}$ [%]	$\bar{\iota}^{\downarrow}$ [%]	$1-\bar{\lambda}^{\downarrow}$ [‰]	no. hibl
Berlin	preprocessed	673.8	13.9	773.2	35.9	63.4	114.7	59.6	42
	local search	499.0	2.4	660.7	30.7	52.2	91.2	21.4	21
	MIP heuristic	490.3	2.0	647.9	30.1	49.5	89.2	18.7	14
Lisbon	preprocessed	416.9	12.8	469.8	24.5	60.0	81.8	25.2	12
	local search	351.1	9.4	422.9	22.1	51.3	65.7	9.8	6
	MIP heuristic	350.8	9.1	422.6	22.1	51.6	65.8	9.9	6
Turin[b]	preprocessed	807.0	350.5	998.3	37.6	63.0	234.4	10.2	14
	local search	806.7	350.7	995.0	37.5	62.7	234.2	10.1	13
	MIP heuristic	692.5	353.8	854.4	32.2	57.1	208.5	4.1	6
Vienna	preprocessed	2071.8	51.5	2257.1	31.3	56.0	191.6	15.9	24
	local search	1955.3	53.3	2183.7	30.3	54.3	175.3	11.7	13
	MIP heuristic	1946.6	53.4	2173.6	30.2	54.5	173.0	10.8	11

[a] "obj": objective function (5.19a); $\lvert A \setminus A^{(C)}\rvert_{\ell}^{\downarrow}$: fraction of uncovered traffic load (lacking E_c or E_c/I_0 coverage); $\sum_i \bar{p}_i^{\downarrow}$: total power consumption; $\bar{L}^{\downarrow}$: average cell load; $\max_i L_i^{\downarrow}$: maximum cell load; $\bar{\iota}^{\downarrow}$: average other-to-own-cell interference ratio; $1-\bar{\lambda}^{\downarrow}$: overall blocking ratio; "no. hibl": number of cells with high expected fraction of blocked traffic ($\mathbb{E}(\lambda_i^{\downarrow}) < 98\,\%$).

[b] All results except for the coverage value are restricted to the 133 sectors in the optimization area depicted in Fig. 5.7(a); Tab. B.7 contains the aggregated performance indicators for the complete network.

terference ratio. Finally, the expected overall blocking ratio "$1-\bar{\lambda}^{\downarrow}$" is complemented by the number "hibl" of cells with an expected percentage of blocked traffic larger than 2 %. A complete account of all expected performance figures including uplink values and coverage is given by Tab. B.7 in the appendix.

Tab. 5.4 has three rows per scenario for the preprocessed configuration, the configuration found by local search, and the network design found by the MIP heuristic. The objective function value "obj" is monotonously decreasing in this order because the preprocessed configuration is input to the local search, whose result is in turn input to the MIP heuristic, and only improving configuration changes are adopted throughout. For Turin, the listed numbers consider the sectors in the optimization area only.

Berlin. Preprocessing produces a sparse network configuration with full coverage, but capacity problems. Because the initial configuration already provides full coverage, coverage maximization was omitted. In the subsequent cost reduction step, 43 out of the 68 site candidates were selected. An additional sector reduction was conducted, leading to 108 cells in total. In the resulting network with uniform azimuth and tilt settings, almost 6 % (59.6 ‰) of the traffic are blocked, and in 42 cells blocking is unacceptably high. The average cell load is at 35.9 %.

The result of the local search procedure demonstrates a significant impact of net-

Figure 5.9: Impact of capacity optimization on grade of service

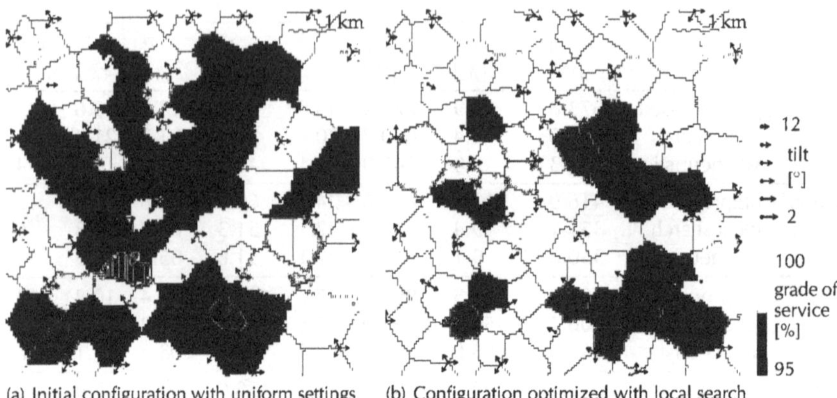

(a) Initial configuration with uniform settings (b) Configuration optimized with local search

work optimization on cell load. The reconfiguration of 85 sectors reduced the objective function value from originally 673.8 W to 499.0 W. This translates into a decrease in the total expected transmit power of about fifteen percent from 773.2 W to 660.7 W and into a decrease in the average cell load by five percentage points. The load of the most highly loaded cell is decreased by eleven percentage points. (The bottleneck cell is different for the two network configurations.)

Furthermore, performance indicators not directly targeted by the optimization objective are improved significantly. The blocking ratio drops from 59.6 ‰ to 21.4 ‰. The number of congested cells is halved from 42 to 21. The preprocessed configuration cannot cover about 1.4 % of the offered traffic due to violation of the E_c/I_0 condition (cf. Tab. B.7), the optimized configuration reduces this fraction to 0.24 %. The improvements are visually represented in the grade of service plots in Fig. 5.9. The dark areas in the initial configuration (Fig. 5.9(a)) indicate cells with high blocking ratios. In the optimized configuration (Fig. 5.9(b)), the situation is alleviated in almost all cells.

The MIP heuristic has found small but noticeable improvements of the local search solution. By modifying 35 sectors, the objective function has dropped another 9 W to 490.3 W. This is reflected in (slightly) better expected network performance: load and other-to-own-cell interference ratio are reduced while coverage and grade of service are improved. The blocking ratio is reduced by more than ten percent (from 21.4 ‰ to 18.7 ‰), so that only 14 cells remain congested. Thirteen reconfigurations have been found by solving MIPs and another 28 adjustments are made by the subsequent local search. The 4-opt MIP heuristic thus enables to escape from a local minimum such that subsequent local search can achieve significant further improvements.

The MIP heuristic appears to be superior to the plain local search in situations that require the simultaneous reconfiguration of several sectors to yield an improvement. The load plot in Fig. 5.10 gives an overview of the results achieved by both methods and shows an example move in detail that could not be found by local search.

Figure 5.10: Additional improvement through the MIP heuristic, Berlin

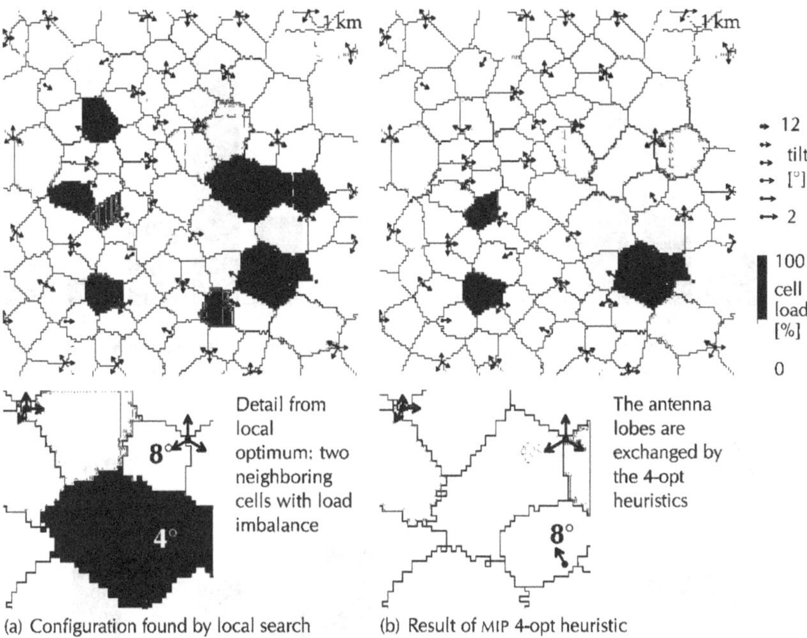

(a) Configuration found by local search (b) Result of MIP 4-opt heuristic

Fig. 5.10(a) shows the local search solution; Fig. 5.10(b) depicts the best solution found by the MIP heuristic. The dark area in the detail corresponds to the cell with the highest load at 4° tilt. This cell has a direct neighbor at 8° tilt, which is almost empty, yet the load cannot be swapped through a sequence of improving steps: Tilting the upper antenna first creates additional interference for the highly loaded cell, so this move is avoided. Tilting the cell at 4° tilt down first, on the other hand, creates a coverage hole in the middle of the detail area. Only the simultaneous reconfiguration of both sectors leads to the improved, more balanced configuration.

An overall increase in downtilt is the chief reason for better performance. The tilt distributions of the initial solution and the best optimized configuration are shown in Fig. 5.11(a). The downtilt could be increased from the initial uniform setting of 6° in most sectors. This has considerably reduced interference. The decoupling between cells is also reflected in the other-to-own-cell interference ratio: the initial value of 114.7 % is comparable to typical values found in literature (e. g., *Nawrocki et al.*, 2006). In the optimized configurations, this is reduced to values around 90 %, which are an indication of a well-optimized network. (Recall that the downlink other-to-own-cell interference ratio values in the generalized coupling equations developed in Sec. 3.3 have to be converted using the average orthogonality for the comparison with values in other literature; see p. 55.)

Figure 5.11: Impact of optimization on downtilt distribution

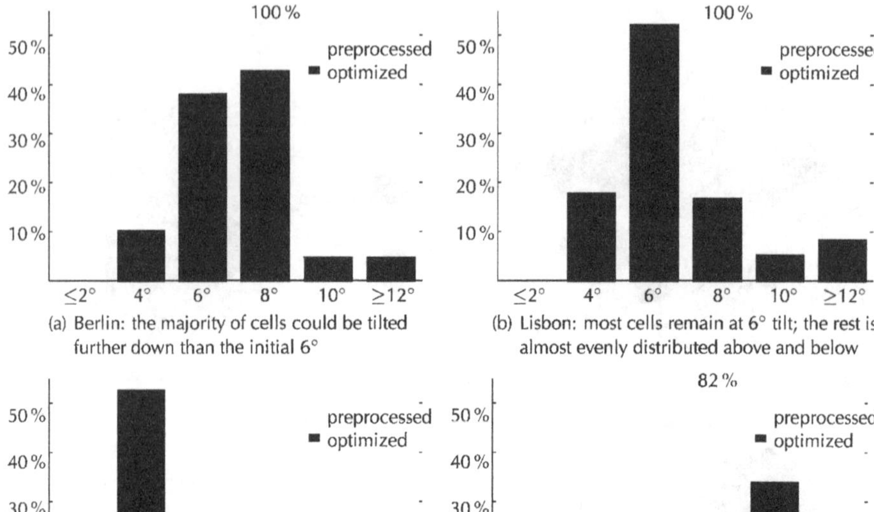

(a) Berlin: the majority of cells could be tilted further down than the initial 6°

(b) Lisbon: most cells remain at 6° tilt; the rest is almost evenly distributed above and below

(c) Turin (urban cells): low values are needed for coverage, can be increased by optimization

(d) Vienna: many cells are tilted to 10° to reduce impact of the antenna's broad main lobe

Lisbon. Also in Lisbon, preprocessing produces a network with (almost) full coverage, but less capacity problems than in the previous case. The initial configuration already has full coverage, so coverage maximization was omitted. In the preprocessing step, 34 out of 60 sites were selected with a total of 96 sectors. Only entire sites were discarded in this scenario. The site reduction fully exploited the allowed reduction in coverage by 0.5 %, leading to a coverage of 99.5 %. In the preprocessed configuration, simulations show that 12 cells have a high blocking ratio and 25.2 ‰ of the traffic is lost on average over the snapshots.

Local search leads to a significantly improved network, which can only be improved marginally by the complex MIP heuristic. The objective function decreases by about 15 % in the first optimization step, which reconfigured 66 sectors. The sum of powers determined in simulation drops about ten percent in the adapted configuration. The blocking ratio falls from more than 2.5 % to under 1 % (9.8 ‰), thereby eliminating 6 out of 12 overloaded cells in the preprocessed configuration. The expected load per cell is depicted for the preprocessed configuration and for the first

Figure 5.12: Impact of capacity optimization on load distribution, Lisbon

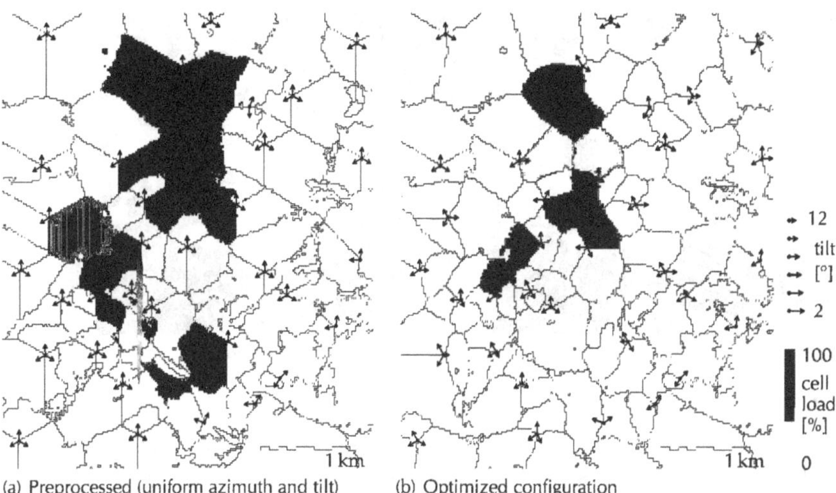

(a) Preprocessed (uniform azimuth and tilt) (b) Optimized configuration

optimization result in Fig. 5.12. The MIP heuristic, however, achieved little gain in the Lisbon scenario: only 5 sectors were modified (1 by the MIP itself causing 4 more changes in local search), reducing total transmit power by 0.3 W, which is a negligible improvement. The blocking ratio and the maximum cell load have even increased slightly, although the change is within the simulation's margin of error.

In general, traffic is served more efficiently in Lisbon than in Berlin because of the advantageous characteristics of the antenna, the higher site density, and the hilly terrain profile. An equivalent amount of traffic is served with considerably lower average load. The scenario has less than half the area of Berlin, so fewer sectors are needed, creating less base load and interference on the pilot channel. (Border effects also play a role, however.) Moreover, with only 81.8 %, the other-to-own-cell interference ratio is significantly lower in the preprocessed configuration than in the best optimized configuration in Berlin. There are two reasons for this: First, the antenna typed used in Lisbon has a vertical opening angle of only 6° compared to 6.5° in Berlin. The difference seems small, but its impact is noticeable since the diffraction angle varies only in a small range. The better focus in the main lobe of the Lisbon antenna is also visible when comparing Figs. A.5(b) and A.5(d). In addition, Lisbon is a hilly city, as Fig. 5.6(a) shows. This provides a natural separation for some cells, as direct neighbors with significant height differences have less overlap.

The tilt distribution in the Lisbon results show that optimization is about making the right decision for the local setting rather than tilting down wherever possible. Unlike in the Berlin case, there is no (strong) bias towards values higher than the initial 6° in the tilt histogram in Fig. 5.11(b). Many sectors have received a tilt of 4°. A well-configured network configuration apparently adapts to local circumstances.

Figure 5.13: Coverage optimization in Turin—indoor and outdoor coverage (dark shades/lighter shades)

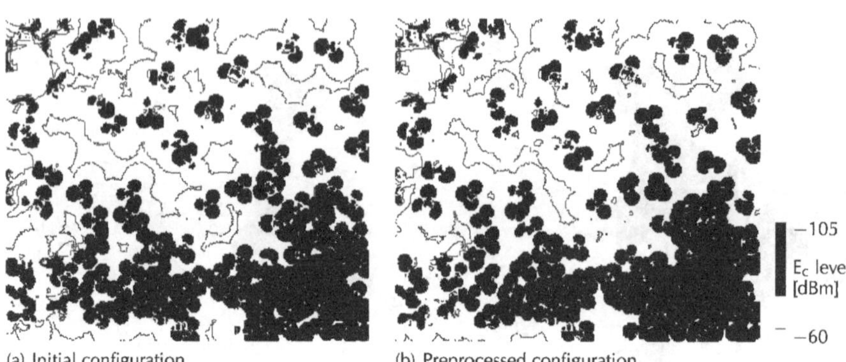

(a) Initial configuration (b) Preprocessed configuration

Turin. Unlike in the previous two cases, ensuring coverage is difficult in the Turin scenario. This has two reasons: First, the density of sites is small. Less than twice as many sites as in Berlin need to serve an area more than five times bigger, and traffic load is distributed all over the planning area. In the rural parts, coverage holes are thus inevitable. Second, there is a comparatively high additional indoor penetration loss of 20 dB considered for 70 % of the users, which applies uniformly throughout the scenario. (For the residual 30 % of the users, an additional loss of 8 dB applies.) We therefore use two notions of spatial coverage, *indoor coverage* and *outdoor coverage*, defined by an adapted threshold value π_{E_c}. The E_c-level plot in Fig. 5.13(a) shows that the initial configuration provides seamless outdoor coverage in the urban area, while there are large coverage holes in the rural areas. Indoor coverage is only provided close to base station sites; even in the urban area there are holes.

We alleviate the situation with a two-step coverage optimization scheme that treats indoor and outdoor coverage separately. First, outdoor coverage is increased by modifying the sector configurations in the rural areas (all sites not included in the polygon indicated in Fig. 5.7(a)). Second, indoor coverage is improved in the urban area with an adapted threshold in (5.17). The resulting indoor and outdoor area coverage levels as well as the covered traffic are listed in Tab. 5.5. Initially, only 27.8 % of the area is covered for indoor users; outdoor coverage is achieved in 81.5 % of the area. This amounts to a total of 60.7 % of the traffic being covered. In the first optimization step, outdoor coverage is increased by five percentage points. Subsequently, indoor coverage is increased by about one percentage point. Note that due to the high traffic intensity in the urban area (cf. Fig. A.2(b)) and the high percentage of indoor users, this small increase has a greater effect on the level of covered traffic than outdoor optimization. After coverage optimization, 62.2 % of the traffic load is covered. The improvement is clearly visible in the coverage plot for the optimized configuration shown in Fig. 5.13(b).

Indoor and outdoor coverage are partly conflicting goals because of the narrow

Table 5.5: Results of two-step coverage optimization in Turin

	cov. area $\lvert A^{(E_c)}\rvert$ [%]		cov. load
	indoor[a]	outdoor[a]	$\lvert A^{(E_c)}\rvert^{\downarrow}_{\ell}$ [%]
initial	27.8	81.5	60.7
outdoor optimization	27.6	86.4	61.1
indoor optimization	28.5	86.2	62.2

[a] Thresholds $\pi_{E_c} = -85\,\text{dBm}$ for outdoor, $\pi_{E_c} = -97\,\text{dBm}$ for indoor

focus of the antennas used in Turin. When in the first optimization step outdoor coverage is improved from 81.5 % to 86.4 %, indoor coverage declines by 0.2 percentage points. Similarly, increasing indoor coverage by one percentage point reduces outdoor coverage. The reason for this counterintuitive behavior is the small opening angle of 4° of the Kathrein 742 213 antenna primarily used in Turin. (The vertical diagram is shown in Fig. A.4(d).) The antenna's main lobe can be either directed at the cell border (for outdoor coverage) or at the cell center (for indoor coverage), but it is not wide enough to serve both areas simultaneously. A two-step approach with different thresholds is therefore necessary (as opposed to a single coverage optimization step) to provide rural and urban areas with the respective appropriate coverage. This phenomenon might be considered an artefact of antenna modeling, but a further investigation is out of our scope.

The local search procedure fails at minimizing capacity in the presence of a hard coverage constraint, because there are hardly any feasible improving moves. We enforced the indoor coverage level in (5.19b). With this constraint, only one sector was found to be amenable to an improving step without sacrificing coverage. This translates to a small improvement in performance as listed in Tab. 5.4. (Recall that the table contains aggregated data for the optimization focus zone only.) The total expected power drops by 3 W, and the maximum expected cell load is reduced by only 0.3 percent points from 63.0 % to 62.7 %. The other performance indicators remain virtually unchanged.

It is not an option to run local search with the outdoor coverage constraint, because too much indoor coverage would be lost. In computational tests, the relaxed outdoor constraint admitted an optimization run that is similarly successful as in the previous scenarios at reducing load. However, in the resulting configuration a significant fraction of indoor coverage is lost, producing a network design that is far worse than the initial configuration with respect to coverage.

The MIP heuristic can play to its strength and leads to significant performance gains because it is able to modify several sectors simultaneously. Coverage areas can thus be swapped among cells, and 54 configuration changes are identified that improve network performance. This time, the final local search step only revises the configuration of 11 sectors. The average cell load is reduced by 15 %; the blocking ratio drops from 10.2 ‰ to 4.1 ‰; only six cells with high blocking remain. The network

Figure 5.14: Capacity optimization Turin: local search can hardly perform any improving steps while maintaining indoor coverage. The 4-opt MIP heuristic improves the network design considerably, reducing cell load by 13 %.
(Only urban cells, which could be modified by capacity optimization, are shown.)

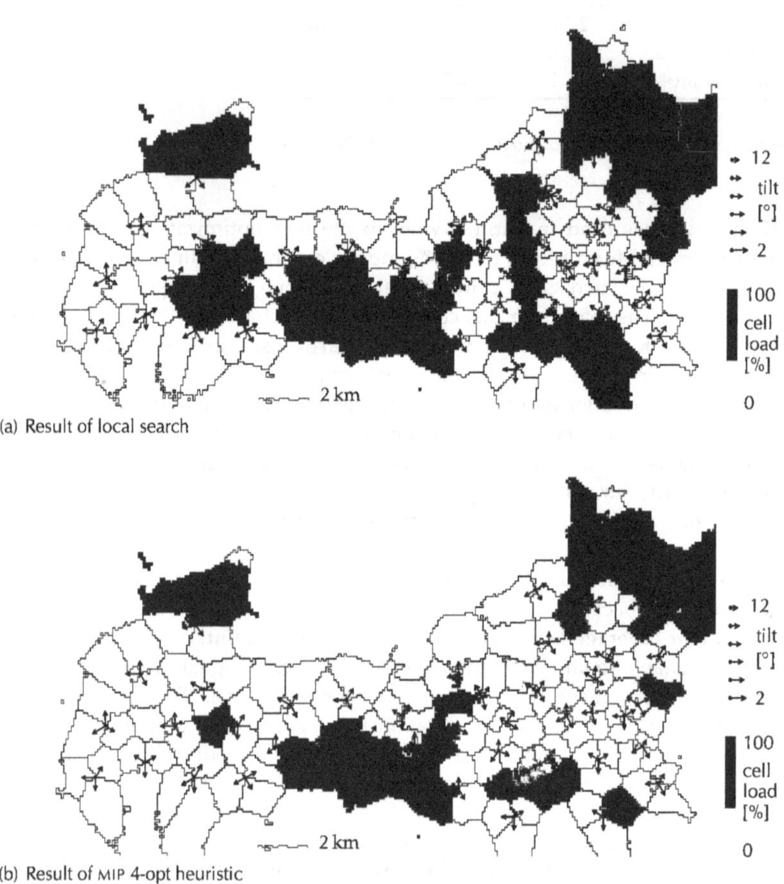

(a) Result of local search

(b) Result of MIP 4-opt heuristic

Figure 5.15: Downlink other-to-own-cell interference ratio in Vienna (part)—optimization interleaves cells

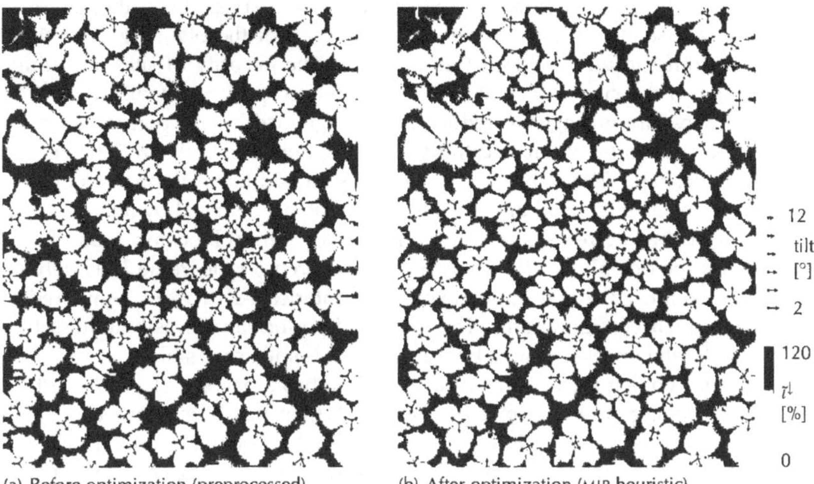

(a) Before optimization (preprocessed) (b) After optimization (MIP heuristic)

configurations before and after optimization and the resulting average cell load values of the urban cells are depicted in Fig. 5.14.

Only sectors with low downtilt values can fulfill the coverage condition, which results in high interference coupling. The tilt histogram in Fig. 5.11(c) shows that more than half of the urban cells have a downtilt of four degree. The average downtilt is considerably lower in Turin than in Berlin and Lisbon. Increasing these values where possible is the key to the improvements obtained in optimization. The low tilt values induce the highest other-to-own-cell interference ratio values observed in all scenarios: Optimization manages to reduce the initial 234.4 % to 208.5 %. Figures in this range are commonly considered "extremely poor network planing" (*Laiho et al.*, 2002). Apparently, the only remedy here is to revise the coverage conditions or the traffic assumptions—or to acquire more sites.

Turin is the only scenario with an uplink bias, and downlink capacity maximization improves uplink performance as well. The detailed results listed in Tab. B.7 show that in all cases, uplink performance is aligned with downlink performance. Average cell load, grade of service, and other-to-own-cell interference ratio all improve with the objective function.

Vienna. The Vienna scenario is the largest dataset. It covers an area of 437 km^2 at a resolution of 50 m and has 628 potential sectors. In preprocessing, an increase of two percentage points in the covered area was obtained by adapting the settings of sites in the rural area depicted in Fig. 5.7(b). The greedy site reduction removed almost half of the sites and sectors. The resulting preprocessed network configuration is overloaded in 24 cells; the average load value of 31.3 % is reasonable.

The scenario is not primarily coverage-driven, so the results of local search and of the MIP heuristic are comparable to the first two scenarios. Even though rural areas with few base station sites make up a large part of the planning area, there is hardly any traffic in these regions (see Fig. A.3). The majority of sites and traffic is concentrated in the city area, which is the focus of capacity maximization. The local search procedure manages to reduce the average load by one percentage point and the blocking ratio by about one third. Only 13 out of 24 cells are left with an elevated blocking rate. The MIP heuristic provides an additional improvement on a smaller scale: The average load and the blocking ratio are reduced by an additional 0.1 %, leading to two more cells remaining below the critical blocking ratio of 2 %. Fig. 5.15 shows the downlink other-to-own-cell interference ratio value $\bar{\iota}^{\downarrow}$ as defined in (3.36). The optimized configuration exposes an interleaved structure compared to the original, regular configurations. The improved performance was bought with a small increase of the uncovered area by 0.2 %, which is still admissible.

The antenna "60 deg sectorized" used in Vienna has a particularly wide vertical lobe with an opening angle of 15° (see also Fig. A.4(f)); this is reflected in a small relative optimization gain and a large other-to-own-cell interference ratio. The broad main lobe leaves little room for improvement given the small range of allowed tilts. In consequence, there is only a relative improvement in average load of under five percent; this is the smallest value for all scenarios. The other-to-own-cell interference ratio, on the other hand, has values between 173 and 192 %, which is almost as large as for Turin—even though an extremely high fraction of sectors is tilted to the maximum value of 10° (see Fig. 5.11(d)) in an apparent attempt at avoiding excessive coupling.

Computational effort. The data in Tab. 5.6 indicate the computational effort of the optimization algorithms. For local search, the total number of evaluations of the objective function and the total running time is stated in columns "# eval" and "time". For the MIP heuristic, we state the total number of MIPs solved in column "MIPs". In addition, the column specifies the number of MIPs that produced solutions improving the objective. The following three columns show the average dimension of the models in terms of rows, columns, and non-zeroes; the numbers include the effect of CPLEX preprocessing. The columns "# eval LS" and "time LS" indicate the number of objective function evaluations and the running time for the local search ; column "time" indicates the total time spent on solving MIPs and the subsequent call to local search for polishing the solution.

The data shows that the local search method is sufficiently fast. For small and medium-sized scenarios, the running time in the range of a few minutes for Berlin to three and a half hours for Turin on a personal computer is acceptable. More than nine hours of running time for the Vienna scenario is comparatively long. The current implementation is, however, not specifically tailored to handle large scenarios. A small investment into streamlining the code and data structures for these situations should reduce the running time. There are few reference values for the number of evaluations; our numbers lie between those reported by *Gerdenitsch* (2004): 80–150 network evaluation for simulated annealing and 150 000 evaluations for a genetic algorithm.

Table 5.6: Computational effort of optimization methods

	local search		MIP heuristic						
	# eval	time[a]	MIPs[b]	∅ rows[c]	∅ cols[c]	∅ nonz.[c]	#eval LS	time LS[a]	time[a]
Berlin	3 915	0:07	58/ 7	50 349	41 688	255 445	2 501	0:04	27:55
Lisbon	8 415	0:30	41/ 1	59 944	35 676	331 482	10 470	0:40	18:05
Turin	29 538	3:32	107/34	22 030	19 026	94 196	32 742	3:45	13:13
Vienna	49 200	9:39	141/18	24 477	20 549	114 761	41 445	8:04	62:22

[a] Total computation time [h:mm], Intel® Core™ 2 Quad CPU, 2.66 GHZ, 3.4 GB RAM
[b] Number of MIPs: total/improving
[c] Average MIP dimension after CPLEX preprocessing

For the MIP heuristic to be competitive, on the other hand, a speed-up is needed. Only the 13 hours spent on the Turin scenario, in which the MIP heuristic was most successful, may be realistic for practical use. In the other cases, the running time is unacceptably high. Most time is spent on solving MIPs. (The size and running time hardly varied across the MIPs in one scenario.) Even though only four sectors are considered, a large number of variables is needed to compute the objective function. Most subproblems, however, did not yield any improvement, so a smarter choice of fewer subproblems should reduce the effort. A complete enumeration of every subproblem, however, would have taken much longer, as the following section reports. In the light of the comparatively small improvements in network performance achieved with the high effort, the method needs to be refined for practical use.

5.5 Analysis of case study results

The computational results for capacity optimization raise some questions regarding the quality of optimization results and the performance of the algorithms. As we could only apply heuristics, the optimality gap of the resulting configurations is unknown. Furthermore, the performance of local search and MIP heuristic varied across scenarios, and the reasons for the respective performance could shed a light on the complexity of the capacity optimization problem.

To find out why local search was largely successful, we enumerate the search space for small example problems and analyze its structure in Sec. 5.5.1. An analysis of the approximation quality of the convex surrogate objective illuminates the behavior of the MIP heuristic in Sec. 5.5.2. We compare the objective function values of the best solutions found to lower bounds and to reasonable benchmark values in Sec. 5.5.3.

5.5.1 Search space structure of enumerable example problems

Local search seems successful in practice, but it might actually get stuck in a local optimum and fail to come close to the best network configuration. To find out if the initial network configuration was particularly bad or if local search performed well,

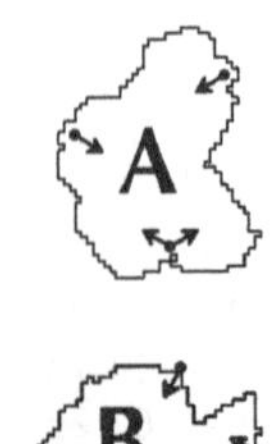

Figure 5.16: Enumerated subproblems "A" and "B", Berlin

we enumerate the complete search space for typical example problems and analyze the structure of the local minima. A similar investigation by *Nawrocki et al.* (2005) suggests that there are many local minima, and that some of them have a significantly higher objective value than the global optimum. This clashes with our impression that local search performs well. We first aim at reproducing their results and then carry the method over to our setting.

We consider the two neighborhoods used by *Nawrocki et al.* (2005) and the one employed in our algorithm. In the "local" neighborhood, configurations only differ in the tilt or azimuth value of a single sector by one step ($\pm 1°$ or $\pm 2°$ tilt, according to the setting; $\pm 30°$ azimuth). In the "diagonal" neighborhood, all sectors are allowed to vary by one step simultaneously. In the "sector" neighborhood, a single sector may be changed arbitrarily. The local neighborhood is always smallest. For the small example problem, the diagonal neighborhood is smaller than the sector neighborhood (because there are more configuration options per cell than modifiable cells). In larger problems, the "diagonal" neighborhood becomes very large. *Nawrocki et al.* (2005) consider the local and diagonal neighborhoods; we use the sector neighborhood.

We consider a synthetic setting and two subproblems for Berlin. The synthetic case is comparable to "Scenario 1" by *Nawrocki et al.*: Two sites with together three antennas of type K 742 264 are positioned at a height of 23 m on a flat plane within 1 000 m of each other; traffic is distributed uniformly in an irregular area. The combined tilt is varied over the integer values in 0–16°. The search space thus consists of $17^3 = 4\,913$ points. The Berlin subproblems were solved by the approximated MIP during the optimization run, but no improving configurations could be found. The two choices "A" and "B" of modifiable sectors (four in each case) are depicted in Fig. 5.16. As there are 15 options per sector, the search space has size $15^4 = 50\,625$.

We enumerated the complete search space and determined the objective function values of all local minima under the three neighborhood definitions. The minima were evaluated for 25 single snapshots separately in the synthetic setting and for the

Table 5.7: Structure of local minima in the enumerated problems

		neighborhood		
		local	diagonal	sector
synthetic scenario, 25 snapshots	no. local minima	56.2	37.1	3.7
	gap $\leq 1\,\%$	4.4	2.9	2.8
	gap $\leq 5\,\%$	22.4	14.6	3.7
	gap $> 5\,\%$	33.9	22.5	0.1
synthetic scenario, expected cpl.	no. local minima	22	16	2
	gap $\leq 1\,\%$	2	1	2
	gap $\leq 5\,\%$	10	7	2
	gap $> 5\,\%$	12	9	0
Berlin Subproblem A	no. local minima	1	1	1
Berlin Subproblem B	no. local minima	1	1	1

expected interference coupling scheme in all three cases. For each Berlin subproblem, complete enumeration took more than ten hours. The number of local minima and the distribution of their objective values are listed in Tab. 5.7. For the synthetic scenario, the table specifies a histogram of the gaps of objective function values compared to the global optimum.

The experiments reproduce the previous results for the local and diagonal neighborhoods. For snapshots, there are on average 56.2 and 37.1 local minima; more than half of them (33.9 and 22.5) have an optimality gap of more than 5 %. For expected coupling estimates, there are 22 and 16 minima, and 12 and 9 of them have an optimality gap exceeding 5 %. The absolute numbers are smaller, but the relative distribution is similar. Also *Nawrocki et al.* observed that averaging has a smoothing effect and reduces the number of minima. In the synthetic cases, less than 10 % of the minima have a gap of less than one percent to the optimum.

For the sector neighborhood and in the Berlin subproblems, the structure collapses. In the synthetic case, few local optima remain, and their objective value is close to the optimum. From the average of 3.7 minima for snapshots, only 0.1 had an objective more than 5 % above the optimum. In the expected coupling case, there is only one non-global minimum, and it has an objective value within 1 % of the global optimum. In the Berlin subproblems, there is only one minimum throughout. The corresponding configuration was already found by local search in our experiments, so it was impossible for the MIP heuristic to improve the objective any further.

Our results suggest that the combinatorial structure of the interference minimization problem is comparatively benign; local search can lead to a solution with little optimality gap in the scenarios considered, if neighborhoods are sufficiently large. The large number of local optima observed in previous research seems due to effects in the synthetic data.

Table 5.8: Approximations of the objective and optimization gain

	correlation with objective			gain [%]	
	linear	convex	combined	MIP	MIP+LS
Berlin	0.664	−0.042	0.867	0.14	1.74
Lisbon	0.955	−0.701	0.173	0.04	0.09
Turin	0.997	−0.720	0.996	12.61	14.16
Vienna	0.968	0.227	0.979	0.32	0.44

5.5.2 Convex approximation of the objective

The performance of the MIP heuristic as compared to local search varied across scenarios. We want to better understand the reasons behind this. The MIP heuristic is a 4-opt scheme with an approximated objective function. Allowing the simultaneous modification of four sectors rather than a only one should always yield an advantage, so we search for the difference in the quality of the approximation. An accurate approximation of the objective function is, however, not necessary for optimization; it is sufficient if the trend is captured. Therefore, we do not evaluate the error of the approximated values (5.16a), but their *correlation* to the actual cell powers (5.8a). For each scenario, we consider all network instances that were found during optimization.

Fig. 5.17 illustrates the relation between the objective function and its convex approximation used in the MIP heuristic. Each dot indicates the approximated and accurate objective function value for an incumbent solution found during MIP solving; the subsequent local search is excluded. The darker dots correspond to those incumbents that improved the current solution and were adopted. In addition, the start solution is indicated. Note that the scales differ in all plots, because the range of average load values is different in each scenario. If the approximation were precise, the points in the plots in Fig. 5.17 would lie on the sector diagonal. Points lying on a straight line (with positive slope) indicate high correlation; the farther the set of points is from forming a line, the smaller is the correlation, and the less useful is the approximation in optimization. In Lisbon, Fig. 5.17(b), the values are not in a clear relation. In Berlin, Fig. 5.17(a), the trend seems better, but the points are still disparate. Vienna already comes close to a linear relationship, and in Turin correlation is high.

The correlation between approximation and objective is related to the gain achieved through optimization. The numbers corresponding to the plots in Fig. 5.17 are listed in column "combined" in Tab. 5.8. The value of 0.173 for Lisbon indicates hardly any correlation, and, as expected, Turin has the highest correlation. In addition, the table states the gain of the MIP heuristic over plain local search. The column "MIP+LS" corresponds to the objective values as listed in Tab. 5.4. In particular, the optimization gain includes the improvements made by the final local search. Column "MIP" singles out the gain achieved by solving the MIPs only. Clearly, this gain reflects the correlation values. Turin has the highest gain with 12.61 % and also the highest correlation; also the gains achieved in the other scenario are ranked according to the

Figure 5.17: Approximated objective vs. accurate objective

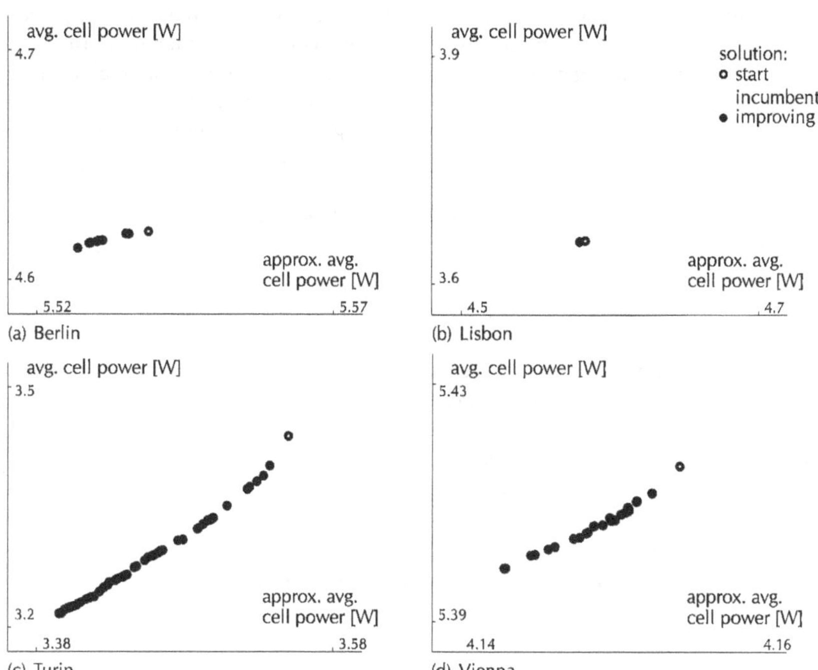

correlation. Concluding that high optimization gain is an effect of high correlation, however, seems infeasible given only the data presented here; we picked the sample of configurations from the solutions produced by the MIP heuristic, which itself depends on the approximation.

Tab. 5.8 also shows how the (purely) convex approximation complements the linear one. Column "linear" contains the correlation of the average transmit power to the average of the linear approximations (5.14a). Column "convex" lists the correlation to the average of the values (5.15a). The linear approximation alone seems reasonable in most scenarios. The convex one alone is not helpful, it is even negatively correlated to the transmit powers in all scenarios but Vienna. If both are combined, however, significantly better estimates are obtained in Berlin and Vienna. In Turin, the precision of the linear estimate is practically left unaltered. Only in Lisbon, where the MIP heuristic achieved very little, adding the convex estimate reduces the correlation.

5.5.3 Bounds on the optimum

For assessing the quality of our solutions, we need to find bounds on the optimum value. For obtaining tight bounds, it is necessary to consider local properties of the input data. This seems basically as hard as optimization itself. As an alternative, we derive two global bounds on the average load factor and the average transmit power, and combine these to a lower bound on the average cell load in each scenario. The bound is based on idealizing assumptions. To put our results into a more realistic perspective, we furthermore calculate a scenario-specific benchmark value that is based on a regular configuration in a homogeneous setting. We use the notation introduced in Secs. 2.4.3 and 4.2.1; a complete list of symbols is given on pp. 169ff.

Perfect inter-cell interference reduction and perfect load balancing. Recall that for a non-empty cell i, the load factor is defined in Sec. 3.3 as $(1+\bar{\iota}_i^{\downarrow})c_{ii}^{\downarrow} \geq 0$. For the sum of the load factors, it holds that

$$\textstyle\sum_{i\in\mathcal{N}}(1+\bar{\iota}_i^{\downarrow})c_{ii}^{\downarrow} \geq \sum_{i\in\mathcal{N}} c_{ii}^{\downarrow} \overset{\text{Lem. 4.1}}{=} \int_{A(\mathbb{E}_c)} \omega(x) T_\ell^{\downarrow}(x) \mathrm{d}x\,; \tag{5.20}$$

If a given area is to be covered, this represents a bound on the average load factor. We call it the *perfect inter-cell interference reduction bound* because it is assumed that inter-cell interference vanishes.

For a given average load factor, on the other hand, it is intuitively clear that minimum total load is obtained by distributing the load evenly among cells, because this reduces intra-cell interference. We can see that this is indeed a bound if pilot powers are uniform with the following lemma:

Lemma 5.2: *The minimum of* $f : [0,1)^n \to \mathbb{R}_+$ *with* $f(\boldsymbol{x}) := \sum_i 1/(1-x_i)$ *subject to* $\sum_{i=1}^n x_i/n = \bar{x} \geq 0$ *is* $\bar{x} \cdot \mathbf{1}$.

Proof. Consider

$$\begin{aligned} \min \quad & f(\boldsymbol{x}) := \textstyle\sum_i 1/(1-x_i)\,, \\ \text{s.t.} \quad & h(\boldsymbol{x}) := \textstyle\sum_i x_i = n\,\bar{x}\,, \\ & \boldsymbol{x} \geq \mathbf{0}\,. \end{aligned}$$

The gradients of f and h are

$$\nabla f(\boldsymbol{x}) = \left((1-x_i)^{-2}\right)_{1\leq i\leq n}, \qquad \nabla h(\boldsymbol{x}) = \mathbf{1}\,.$$

Because f is convex and h is linear, linear dependency of their gradients is a necessary and sufficient condition for optimality (e. g., *Avriel*, 1976, Theorem 4.39). This can obviously only be the case if $x_i = \bar{x}$ for all i. □

Suppose that pilot power is set uniform throughout the network. Then, according to Lemma 3.9, the convex objective f represents the objective for optimization, if we substitute the cell load factors for the values x_i. The lemma hence bounds our capacity objective function from below for a given load factor. Because the bound assumes evenly distributed load among cells, we call it the *perfect-load-balancing bound*.

Under the assumption that a planning area A is covered completely, we can combine both bounds to the following bound on the objective function:

$$\mathbf{1}^t \sum_{k=0}^{\infty} (C^{\downarrow})^k p^{(c)} \geq \frac{n p^{(c)}}{1 - \int_A \omega(x) T_\ell^{\downarrow}(x) \mathrm{d}x / n}\,. \tag{5.21}$$

Uniform pilot power is given in our scenarios, but complete coverage only in Berlin and Lisbon. For the other two, we cropped the scenario to the area covered by the best solution for obtaining a comparison.

Regular benchmark configuration. It is unlikely that any feasible network configuration can avoid inter-cell interference altogether. The lower bound is therefore supposedly loose, but a better bound is not at hand. The problem is that inter-cell interference occurs predominantly at cell borders, but it is impossible to locally pinpoint the cell borders. With the available degrees of freedom, virtually any given point may end up at a cell center or border, depending on the choice of configurations.

For a benchmark value including inter-cell interference, we compare our optimized solutions to the classical idealized hexagonal configuration. For our case of horizontal opening angles around 60° and three-sectorized sites, we adopt the typical assumptions that all cell shapes are hexagonal; that the three cells at a site form a "clover leaf" (three non-overlapping hexagons with a common edge between any two); that the sites are placed on a triangular grid such that the cells are interleaved; and that all antennas are configured uniformly. This configuration is considered optimal by engineers (*Lee & Miller*, 1998), although no formal notion or proof of optimality (considering interference) seem to be available. The hexagonal cell shapes stem from the simple case of omnidirectional antennas with circular coverage area; their "optimality" is plausible, because with the hexagonal structure the area is covered with least overlap (*Fejes Tóth*, 1953).

We parameterize the regular setting in a scenario-specific way. We distribute sites with the same site density, a homogeneous traffic amounting to the same normalized load per area, and the scenario's antenna specification. All antennas are located at the scenario's average antenna height. The specific values for all parameters are contained in Tab. A.4 in the appendix. From all available tilt values, the one for which full coverage was ensured with least average load was selected and assigned to all antennas throughout the scenario. The selected tilts and the resulting average other-to-own-cell interference ratio are listed in Tab. B.6 in the appendix.

The benchmark configuration contains more information on the scenario than only the pilot power and the total normalized traffic considered in (5.21). In particular, the precise antenna characteristics are used, which we have seen to make a difference in performance and optimization gain. In this sense, the benchmark values are more informative, but they represent no bound on average cell load.

Comparison to bounds. Fig. 5.18 graphically relates the performance of the preprocessed and optimized configuration to the bounds and the benchmark value. It depicts the average load factor vs. the cell load for the two network designs and the benchmark configuration. The vertical lines indicate the perfect inter-cell interference reduction bounds (5.20) on the average load factor, and the convex graphs represent the perfect-load-balancing bound. The perfect load balancing graphs are identical for all scenarios except Vienna (Fig. 5.18(d)), because in Vienna the pilot power is set to 33 dBm, as opposed to 30 dBm in the other three cases. The lower bound (5.21) on cell load corresponds to the intersection of both lines. The benchmark configuration has

Figure 5.18: Average load factor vs. cell load—bounds, benchmark configuration, and instances

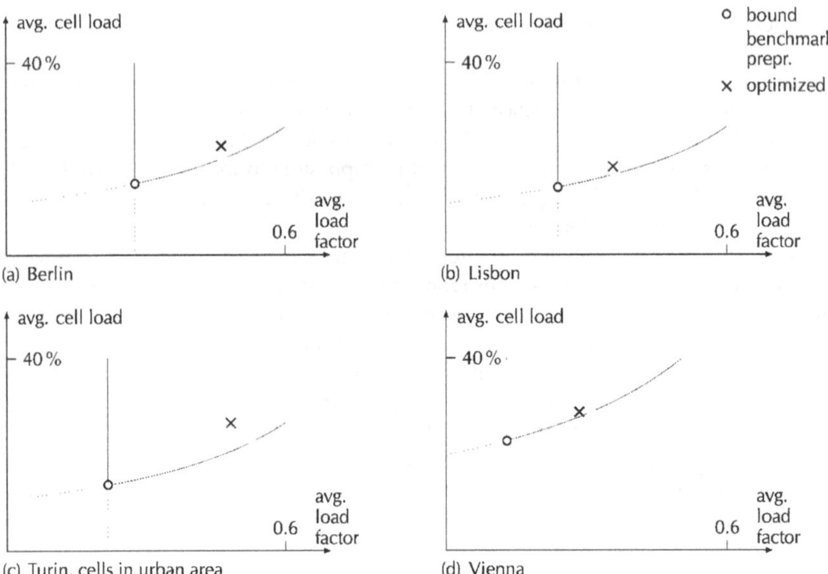

a perfect load balancing by construction, so it is located on the perfect-load-balancing graph. The load factors and cell loads of the optimized solution and the benchmark configuration, as well as the bounds are also listed in Tab. 5.9.

The graphs show that the preprocessed configurations have more inter-cell interference than the benchmark values, and optimization reduces the average load factor to a value that is comparable to the benchmark configuration. Perfect suppression of inter-cell interference, however, remains out of reach. In the case of Turin (Fig. 5.18(c)), the optimized average load factor falls considerably below the benchmark value; this is because the hard coverage condition could only be fulfilled with a uniform tilt of 4° (see Tab. B.6), for which interference is high.

In all cases except Turin, the optimized configuration shows an improved load balancing, and in Fig. 5.18 the optimized configuration lies closer to the perfect load balancing bound than the preprocessed configuration. In Turin, on the other hand, inter-cell interference is apparently reduced by a configuration with less evenly distributed load. This phenomenon is also visible when comparing the load plots. The load distributions in Figs. 5.10(b) and 5.12(b) appear more balanced than before optimization (Figs. 5.10(a) and 5.12(a)). In contrast, the optimized configuration in Turin (Fig. 5.14(b)) is less balanced than before optimization 5.14(a).

The last two columns of Tab. 5.9 compare the average cell load of the optimized configuration to the bound and the benchmark value. As expected, the lower bound is not tight; the gap between the optimized cell load and the bound (5.21) ranges

Table 5.9: Bounds and benchmark values for network planning

	average load factor			average cell load [%]			best gap [%]	
	bound	bench.[a]	best[b]	bound	bench.[a]	best[b]	to bound	to bench.
Berlin	0.28	0.47	0.47	14.8	20.0	22.7	34.7	12.0
Lisbon	0.24	0.36	0.36	14.0	16.7	18.3	23.4	8.4
Turin	0.22	0.53	0.48	13.7	22.7	26.0	47.2	13.0
Vienna	0.13	0.31	0.28	22.8	28.7	28.8	20.8	0.1

[a] benchmark value (performance of best regular cell configuration for homogeneous traffic)
[b] best instance found during optimization

between 20 % and 50 %. From these numbers, we cannot be sure that the results of our optimization come close to the optimum. The gap to the benchmark value is smaller. It lies in the range of 0.1 % for Vienna to 13 % for Turin. The average load factors of the optimized configurations, however, are reduced to the level of the benchmark configuration in Berlin and Lisbon, and even below the benchmark value in Turin and Vienna. The potential of inter-cell interference reduction seems to be exhausted, and the gap is basically due to unevenly distributed load. Because all scenarios have an irregular site structure and inhomogeneous traffic, however, load imbalance is inevitable. The benchmark values thus indicate that a significant untouched potential for improvement is at least unlikely to exist.

5.6 Conclusions on performance optimization

Network performance optimization is a multicriteria, stochastic, combinatorial optimization problem. The task is to select base station sites and find antenna and cell configurations for best expected network capacity, coverage, and cost.

The tradeoff between coverage and cost has been investigated extensively in previous work; usually, either minimum set covering or maximum coverage location problems have been solved. The structure of these problems is well understood, and the objectives can be reasonably evaluated in deterministic models. Capacity optimization, however, needs to consider interference coupling effects among cells, and it is not clear how to do this. The optimization models proposed so far either contain a coarse simplification of interference coupling, or they use an evaluation oracle as an accurate but intransparent notion of coupling. The structure of capacity optimization in interference-limited systems is hence insufficiently researched so far.

We contribute a new model for optimizing expected network capacity, which is based on the system model developed in the previous chapters. The objective is formulated in terms of the expected coupling matrix and aims at minimizing interference. It preserves the nonlinear aspect of interference coupling and allows to analyze the related effects. We propose to solve the model using local search and, in addition, develop a MIP model that approximates the nonlinear objective and is used in a 4-opt heuristic. Model and solution algorithms are tested in four case studies on realistic planning data; to the best of our knowledge, our experiments are more extensive than in any previous contribution.

The experiments demonstrate that the new optimization model is sensible and the proposed algorithms are largely efficient. Reducing the deterministic objective is reflected in the results of detailed simulation: expected cell load, grade of service, and blocking are improved. Moreover, the largest share of the optimization gain is obtained in less than ten hours of computation on high-end PCs even in large scenarios, so the methods are usable in practice. Our methods are heuristic, but the comparison to idealized benchmark solutions suggests that significantly better configurations are unlikely to exist.

The case studies and additional experiments give insight into the practical complexity of the problem and the performance of the algorithms we proposed. Local search has delivered more than 90 % of the improvement in three out of four cases. For small example problems, a complete enumeration of the search space revealed that its structure is simple and the descent method already found the optimum solution. This suggests that also in the large problems, the solutions of local search are good. The MIP heuristic finds better solutions and shows that there is still optimization potential. The additional gain, however, is small, and hardly justifies the increased effort. The Turin case forms an exception, because hard coverage constraints fragment the search space, so only the MIP heuristic delivered significant improvement.

Our findings are good news for practice, because they demonstrate that with a proper model, efficient radio network planning is possible. We have furthermore presented new methods that produce efficient networks in practical settings. For further development, exact algorithms remain a challenge. Our analysis has shown that the nonlinear aspect of interference coupling is essential, and nonlinearity still eludes the capabilites of exact algorithms for most problems. The best candidates for better optimization algorithms are presumably refined heuristics. The larger neighborhood of our MIP heuristic seems to be primarily responsible for the improvements over local search. Algorithms changing more than one sector at a time should therefore be researched further, and structural information such as the derivative of the objective function should be integrated.

Things to remember: Optimization

- The degrees of freedom for planning based on a static model are base station positioning, antenna configuration, and pilot power. The problem is multicriterial with the three objectives capacity, coverage, and cost.
- We specify a new optimization model for maximizing cell capacity over all feasible network designs z by minimizing interference:

$$\begin{aligned} \min \quad & \textstyle\sum_{k=0}^{\infty} \big(C^{\downarrow}(z)\big)^k \big(p^{(c)} + p^{(\eta)}(z)\big) \\ \text{s.t.} \quad & C^{\downarrow}(z) \text{ is the expected coupling matrix for } z \\ & p^{(\eta)}(z) \text{ is the expected noise load for } z \\ & z \in \mathcal{F} \end{aligned}$$

  The capacity model is combined through ϵ-constraint scalarization with known models for cost and coverage.
- We develop a mixed integer linear program that approximates the objective. It is used in a 4-opt heuristic as an alternative to local search with a 1-opt neighborhood.
- In four case studies on realistic scenarios, we use the new model to minimize interference under coverage constraints. The experiments show:
  - Our capacity optimization model has significant impact on the expected performance indicators as determined by accurate simulation.
  - Simple local search produces good solutions in at most a few hours of computation time for large scenarios. The MIP heuristic improves the results because it considers larger neighborhoods, but it takes considerably longer.
- Analyses of the case study scenarios show:
  - Descent methods maximize capacity well, because the search space has a simple structure if ensuring coverage is easy.
  - The nonlinear aspect of interference coupling is essential for optimization.
  - There is a gap of up to 50 % to the best known lower bounds, but the optimization results come close to reasonable benchmark configurations.

Conclusion

The complexity of UMTS technology has left operators struggling to plan and run their radio networks efficiently. Because the radio interface adapts continuously to current traffic and conditions, there is no simple formula to determine the performance of a network configuration. Already a rough dimensioning requires detailed information on user location, traffic characteristics, channel models, and equipment. Only time-consuming simulation can give a full account of network performance. In the initial, coverage-driven deployment phase, operators made do with adapted recipes for second generation technologies and obtained reasonable network designs. In the current period of increasingly competitive markets and growing demand, however, operators need to push the envelope of network performance—but how to bridge the gap between reasonable and highly efficient network planning is not satisfactorily resolved yet.

This thesis contributes models, methods, and insights that help mastering the complexity of UMTS radio network planning. In particular, the new phenomenon of interference coupling has so far not been considered sufficiently in optimization. Our approach focuses consequently on the *expected coupling matrix* for grasping interference coupling and bases capacity optimization on this matrix. We obtain this new paradigm by distilling the essence of the classical evaluation model into a concise representation of relations between radio cells. Realistic case studies demonstrate that the new approach brings an effective fine planning of radio networks within reach.

The basis of our contribution is a new *system model* that replaces classical static simulation with a concise, deterministic first-order approximation of expected network performance. It is developed in two steps: First, we generalize previous modeling approaches to the interference-coupling complementarity systems. These systems describe all relevant aspects of network performance at cell level as a function of the coupling matrix, which was previously only used for calculating cell power levels.

Second, we use the mean coupling matrix as a representative to obtain estimates for expected performance. The resulting model simplifies the impact of individual users and stochastics and does not require simulation, yet computational tests demonstrate that it produces an informative top-level perspective on radio networks.

Based on the system model, our new *capacity optimization model* provides the missing link in radio network planning for UMTS. To complement known models for cost and coverage, we formulate an objective for interference reduction in terms of the expected coupling matrix. In contrast to previous work, the nonlinearity of coupling is explicitly retained. This has turned out essential for fine planning. Optimizing our simplified objective under coverage constraints improves network capacity in all aspects that detailed static simulation evaluates. With the proper model, the problem is even benign, and already simple search methods are effective. For squeezing out the last percents in capacity, we develop new mathematical optimization schemes. The quality of the optimized configurations is assessed by new bounds on interference minimization. A faithful notion of network capacity and of the structure of the capacity optimization problem is thus introduced into automatic radio network planning.

All our methods are geared to practical usage. The data models and detailed structures from practice are not only used in the case studies, but also considered throughout the model development. All algorithms run efficiently on standard computing equipment. Modeling decisions are made transparent, and we provide a modular optimization toolbox that is adaptable to specific practical use cases.

Some questions remain open, and there is potential for extension and generalization. Within the context of the new optimization model, the challenge of finding proven optimal solutions still stands. It may be taken on by refining our methods and ideas for solution heuristics, by conceiving new algorithms, and by finding tighter lower bounds. Furthermore, the optimization framework may be extended to consider aspects that we have disregarded, like uplink performance and pilot power planning. In general, our techniques apply to the modeling and planning of arbitrary interference-limited radio system. Recent work has shown how our methods can be carried over to HSPA extensions of W-CDMA (*Geerdes*, 2007) and to OFDM (*Majewski et al.*, 2007).

Appendices

Data for network planning

Table A.1: Traffic scaling factors against original data for different traffic intensities and service mixes

		traffic intensity		
	service mix	low	medium	high
Berlin	Speech Only	3.2051	4.2735	5.3419
	Video Only	4.1768	5.5690	6.9613
	All Services	0.6000	0.8000	1.0000
Lisbon	Speech Only	1.7862	2.7403	4.3845
	Video Only	2.2833	3.4596	5.5354
	All Services	0.3259	0.4938	0.7901

Table A.2: Additional parameters

parameter		unit	value
maximum load DL	$\bar{p}^{\downarrow}_{\max} / p^{\downarrow}_{\max}$	%	70
maximum load UL		%	50
minimum E_c level	π_{E_c}	[dBm]	−105
minimum E_c/I_0 level	μ_{E_c/I_0}	[dB]	−15
common pilot power	$p^{(P)}$	[dBm]	30
slow fading standard deviation		[dB]	8

Table A.3: Antenna models used for case studies

model	scenario	elec. tilt	opening angle horizontal	opening angle vertical	gain [dBi][b]
Kathrein 742 212	Berlin	0–8°	63°	6.5°	18.0
Kathrein 742 265	Lisbon	0–6°	65°	6.0°	18.5
Kathrein 741 784	Turin[a]	0–8°	60°	6.0°	18.0
Kathrein 742 213	Turin[a]	0–6°	65°	4.0°	19.5
Kathrein 742 271	Turin[a]	0–6°	56°	7.0°	18.5
"60 deg sector"	Vienna	0°	58°	15.0°	15.0

[a] Type 742 213 is used predominantly (302 out of 335 sectors)

[b] Amplification of the signal in direction of the main lobe as compared to a perfectly isotropic radiator (in dB)

Figure A.1: Key geographical data, Berlin scenario

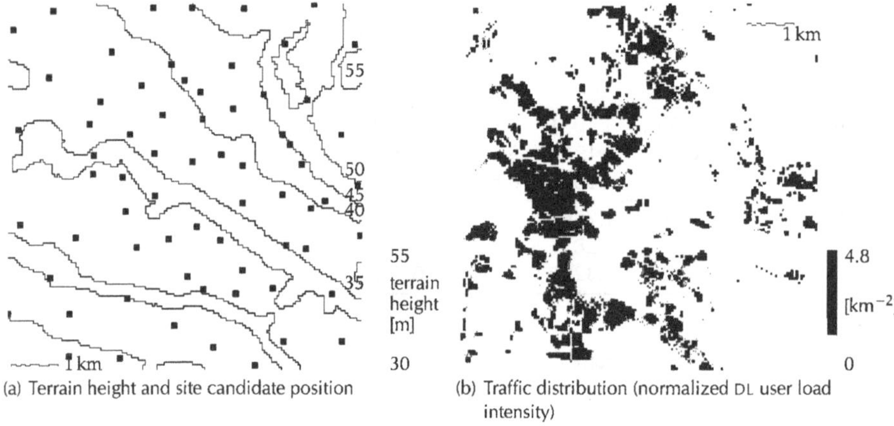

(a) Terrain height and site candidate position

(b) Traffic distribution (normalized DL user load intensity)

Figure A.2: Key geographical data, Turin scenario

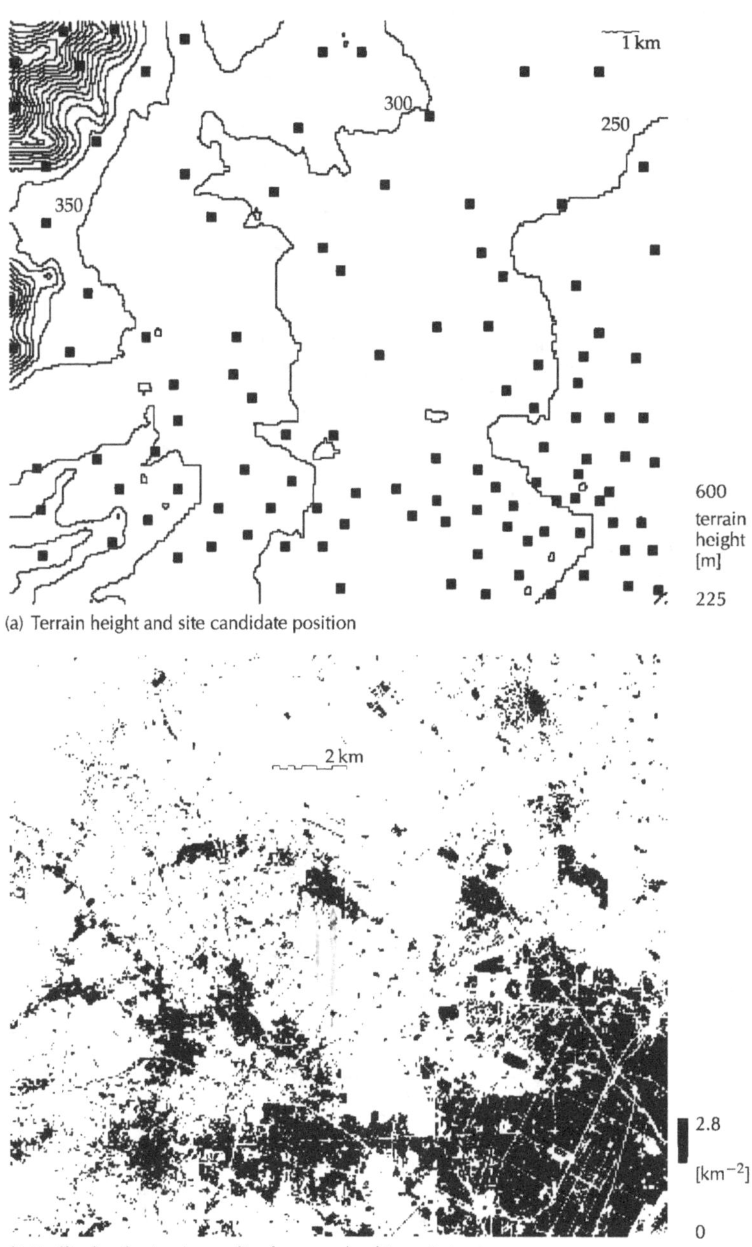

(a) Terrain height and site candidate position

(b) Traffic distribution (normalized DL user load intensity)

Figure A.3: Key geographical data, Vienna scenario

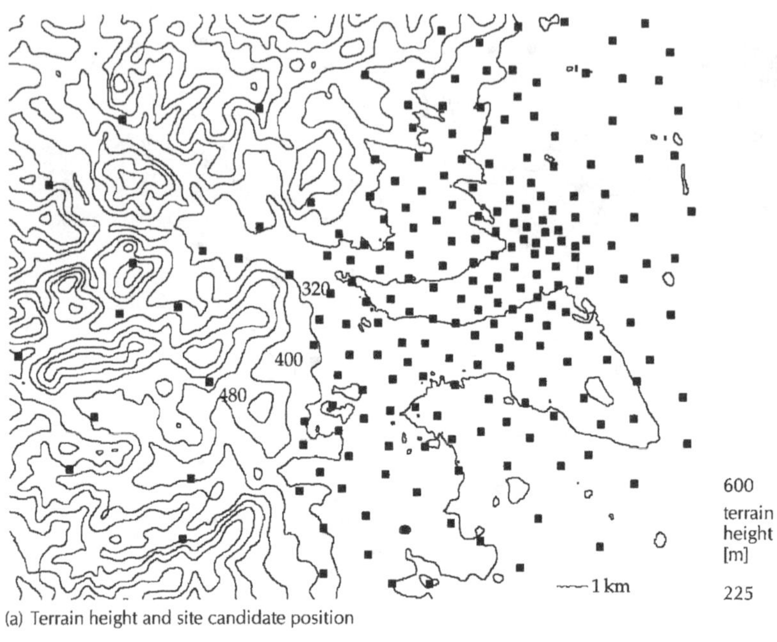

(a) Terrain height and site candidate position

(b) Traffic distribution (normalized DL user load intensity)

Table A.4: Scenario data

	height [m]		extension [km]		area
	min	max	x	y	[km^2]
Berlin	32.0	55.0	7.5	7.5	56.25
Lisbon	0.0	106.0	4.2	5.0	21.00
Turin	222.0	1 164.0	17.9	15.4	274.00
Vienna	224.0	988.0	23.0	19.0	437.00

	no. of	no. of	sector	antenna height [m]		
	sites	sectors	[°]	min	avg	max
Berlin	68	204	90	23.0	35.7	70.0
Lisbon	60	164	0	7.0	24.5	39.0
Turin	111	335	30	20.0	30.1	48.0
Vienna	211	628	30	27.0	29.5	32.0

	resolu-tion [m]	avg. ortho-gonality	total user load	avg. user load intensity [km^{-2}]
Berlin	50	0.414	51.64	0.92
Lisbon	20	0.339	52.45	2.50
Turin	50	0.633	96.40	0.35
Vienna	25	0.600	97.43	0.22

Table A.5: Service parameters and traffic mix for MORANS scenarios (Turin and Vienna)

		BLER	data rate	activity	CIR target [dB][a]			share[b]
		[%]	[kbps]	factor [%]	min	avg	max	[%]
Turin	Voice	1	12.2	0.5	−15.2	−14.8	−14.1	48.7
	Video	1	64.0	1.0	−11.7	−11.2	−10.4	44.4
	Data	1	32.0	1.0	−14.8	−14.3	−13.6	6.9
Vienna	Voice	1	12.2	0.5	−15.2	−13.7	−12.7	48.4
	Data 1	10	128.0	1.0	−9.0	−7.6	−6.6	28.1
	Data 2	1	64.0	1.0	−11.7	−10.0	−8.9	23.4

[a] The CIR target depends on the user's velocity
[b] Percentage of overall traffic load

Table A.6: Service parameters and traffic mix for MOMENTUM scenarios (Berlin and Lisbon)

	BLER [%]	data rate[a] [kbps]	activity factor[b]			CIR target [dB][c]			share [%][d]	
			min	avg	max	min	avg	max	Berlin	Lisbon
Voice	1	12.2	0.50	0.50	0.50	−17.5	−17.3	−16.8	18.7	18.2
Video	1	64.0	1.00	1.00	1.00	−12.1	−12.1	−12.1	14.3	14.4
Streaming	1	64.0–128.0	1.00	1.00	1.00	−12.1	−9.8	−9.6	59.3	59.5
EMail	10	32.0–128.0	0.83	0.91	0.95	−15.6	−12.3	−8.4	0.1	0.1
Download	10	64.0–384.0	0.44	0.88	0.90	−12.5	−12.1	−6.5	6.9	7.1
MMS	10	32.0–128.0	0.82	0.90	0.95	−15.6	−11.8	−8.4	0.3	0.2
WWW	10	32.0–384.0	0.36	0.73	0.87	−15.6	−11.5	−6.5	0.5	0.5

[a] Data users may be assigned different bearers according to RRM policy and availability

[b] For data services, the activity factor is adapted to the assigned bearer; for a higher connection speed, less activity is assumed

[c] The CIR target depends on the assigned bearer and on the user's velocity

[d] Percentage of overall traffic load

Figure A.4: Antenna diagrams of most frequently used antennas

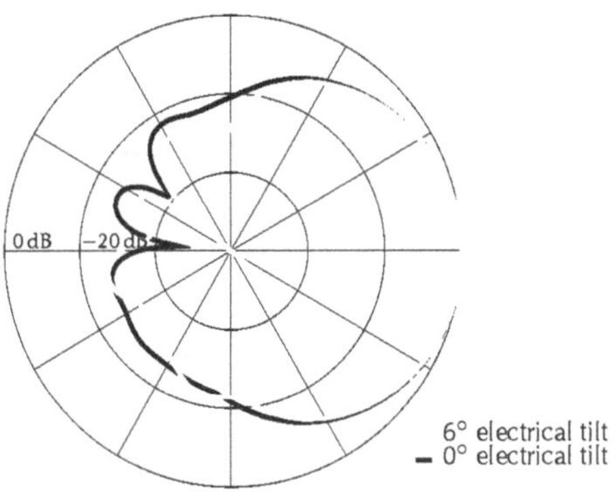

(a) Kathrein 742 265, horizontal plane

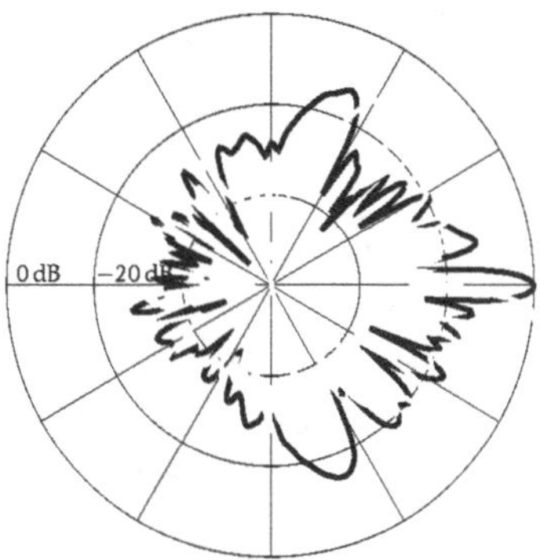

(b) Kathrein 742 265, vertical plane

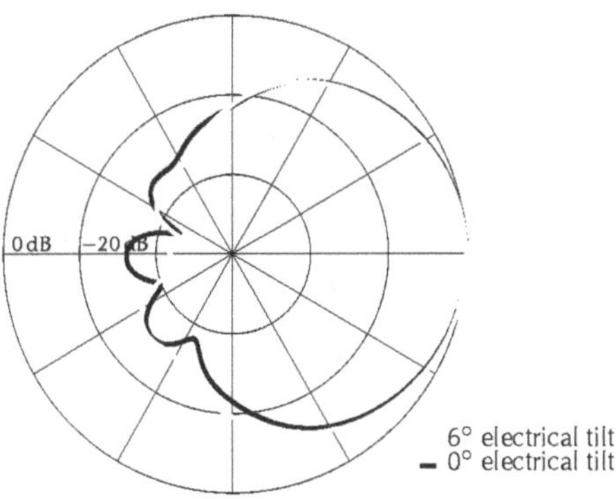

(c) Kathrein 742 213, horizontal plane

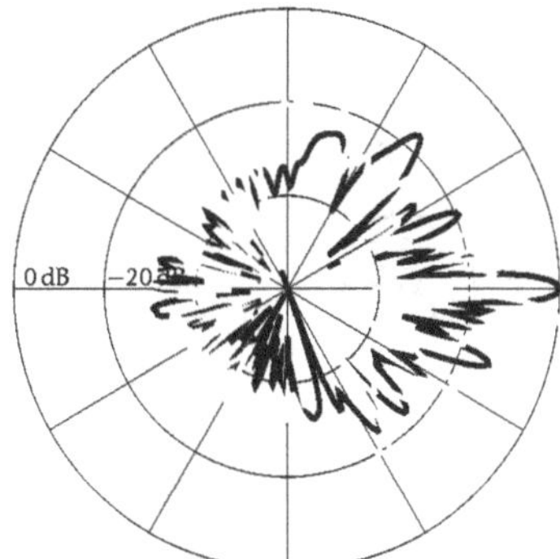

(d) Kathrein 742 213, vertical plane

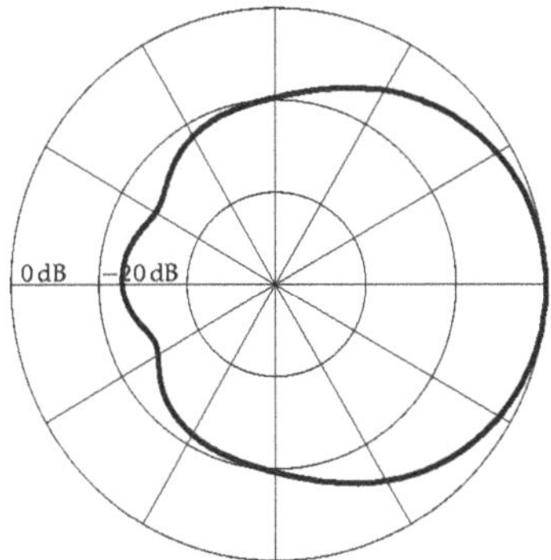

(e) "60 deg sector", horizontal plane

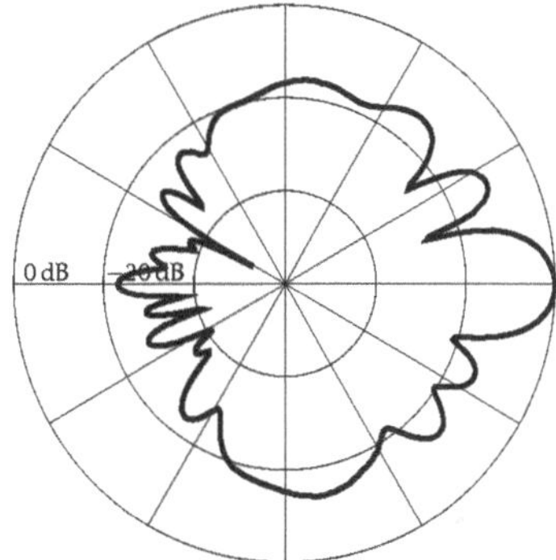

(f) "60 deg sector", vertical plane

Figure A.5: Smoothed antenna diagrams used in the MOMENTUM scenarios

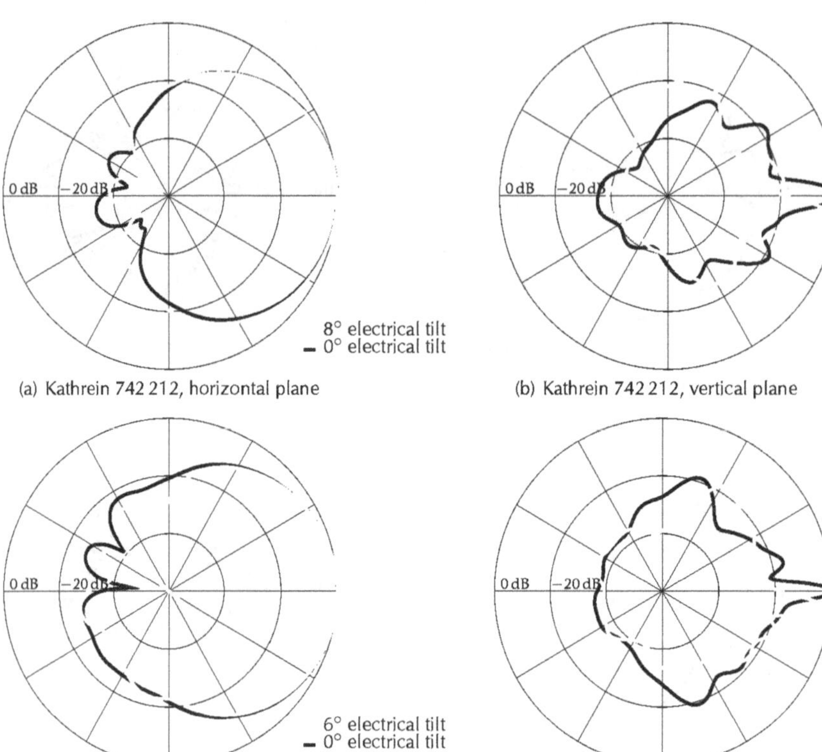

(a) Kathrein 742 212, horizontal plane

(b) Kathrein 742 212, vertical plane

(c) Kathrein 742 265, horizontal plane

(d) Kathrein 742 265, vertical plane

Additional details on computational results

Table B.1: Estimation of other-to-own-cell interference ratio, network "Berlin opt."

		no shadowing			shadowing		
intensity		speech	video	all	speech	video	all
low	mean rel. error	0.062	0.104	0.111	0.332	0.342	0.363
	1−correlation	0.005	0.041	0.011	0.146	0.167	0.170
medium	mean rel. error	0.031	0.078	0.073	0.305	0.286	0.297
	1−correlation	0.002	0.025	0.012	0.188	0.200	0.246
high	mean rel. error	0.029	0.062	0.062	0.286	0.279	0.287
	1−correlation	0.001	0.012	0.011	0.180	0.225	0.277

Table B.2: Approximation of coverage probability

		Speech			Video			All		
intensity	network	opt. thr[a]	opt. agr[b]	cm. agr[c]	opt. thr[a]	opt. agr[b]	cm. agr[c]	opt. thr[a]	opt. agr[b]	cm. agr[c]
Low	reg.	−14.4	0.997	0.984	−14.0	0.995	0.984	−13.7	0.994	0.984
	opt. 1	−14.5	0.999	0.988	−14.7	0.995	0.983	−15.0	0.994	0.982
	opt. 2	−14.4	0.995	0.972	−14.2	0.992	0.974	−14.2	0.992	0.972
	opt. 3	−14.3	0.979	0.948	−14.5	0.984	0.957	−14.3	0.984	0.959
Medium	reg.	−13.1	0.955	0.952	−13.2	0.956	0.953	−13.2	0.958	0.954
	opt. 1	−12.8	0.917	0.916	−12.9	0.930	0.928	−12.9	0.934	0.930
	opt. 2	−12.6	0.899	0.899	−12.6	0.907	0.907	−12.7	0.908	0.908
	opt. 3	−12.4	0.897	0.894	−12.5	0.904	0.901	−12.6	0.906	0.904
High	reg.	−12.6	0.922	0.921	−12.5	0.924	0.923	−12.5	0.924	0.924
	opt. 1	−12.3	0.878	0.869	−12.4	0.885	0.883	−12.4	0.887	0.886
	opt. 2	−12.1	0.884	0.861	−12.3	0.884	0.867	−12.2	0.884	0.869
	opt. 3	−12.1	0.885	0.857	−12.1	0.884	0.862	−12.0	0.885	0.864

[a] Threshold for best agreement between simulation result and approximation [dB]
[b/c] Fraction of correctly predicted pixels for optimal threshold/for common threshold of 12.7 dB

Table B.3: Effect of refined estimation on the prediction accuracy for grade of service

		exp.-coupling estimate			refined estimate		
intensity		speech	video	all	speech	video	all
low	max underestimation	0.000	0.000	0.000	0.000	0.000	0.000
	max overestimation	0.001	0.033	0.042	0.001	0.010	0.012
	max rel. error	0.001	0.035	0.044	0.001	0.010	0.013
	mean rel. error	0.000	0.002	0.003	0.000	0.001	0.001
	1−correlation	n/a*	n/a*	n/a*	0.002	0.014	0.016
medium	max underestimation	0.015	0.070	0.073	0.006	0.046	0.049
	max overestimation	0.045	0.076	0.076	0.007	0.004	0.005
	max rel. error	0.048	0.089	0.093	0.008	0.059	0.063
	mean rel. error	0.005	0.017	0.020	0.001	0.004	0.005
	1−correlation	0.055	0.238	0.258	0.001	0.002	0.003
high	max underestimation	0.016	0.063	0.072	0.008	0.032	0.039
	max overestimation	0.037	0.079	0.079	0.006	0.003	0.004
	max rel. error	0.039	0.087	0.098	0.012	0.045	0.054
	mean rel. error	0.006	0.023	0.026	0.001	0.009	0.010
	1−correlation	0.003	0.026	0.030	0.000	0.001	0.001

* All estimates are invariant at 100 %, so correlation is undefined

Figure B.1: Grade-of-service estimates, medium traffic intensity

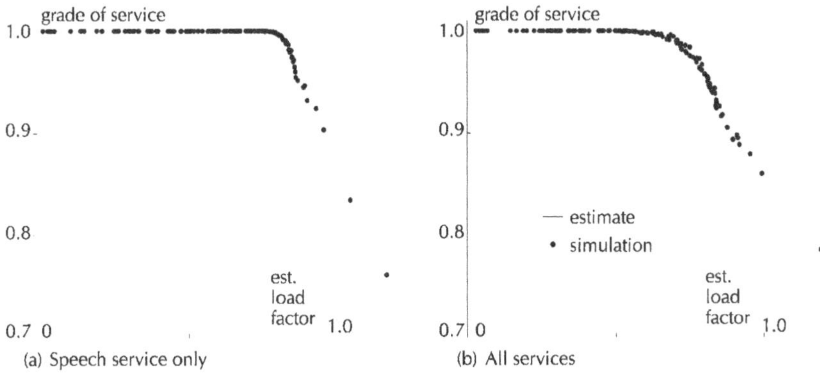

(a) Speech service only

(b) All services

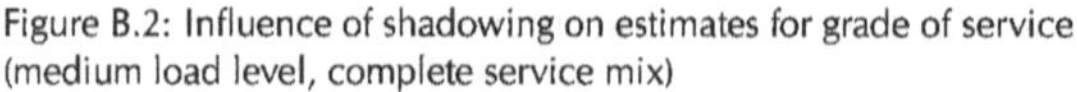

Figure B.2: Influence of shadowing on estimates for grade of service (medium load level, complete service mix)

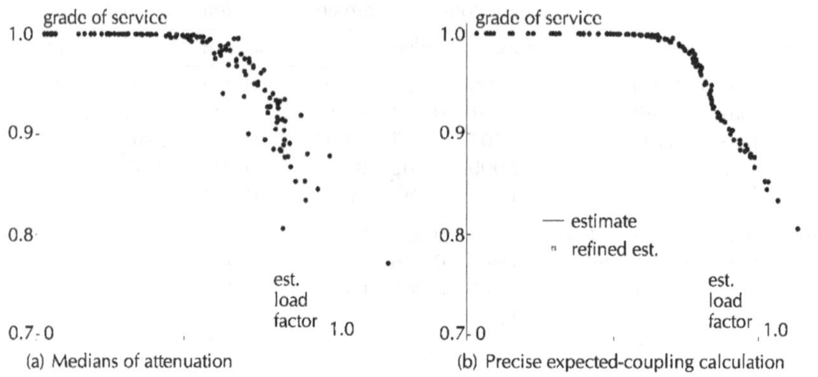

(a) Medians of attenuation

(b) Precise expected-coupling calculation

Table B.4: Performance of discrete load-control methods for Berlin network: avg. transmit power

		speech service				all services			
intensity	method	% hi[a]	⌀ ov[b]	⌀ und[c]	⌀ all[d]	% hi[a]	⌀ ov[b]	⌀ und[c]	⌀ all[d]
low	Rnd. Act.	46.43	0.447	0.399	0.000	62.59	3.171	2.588	0.108
	Rnd. Order	14.29	0.198	0.233	0.000	4.94	0.606	1.999	0.071
	Knapsack	35.71	0.091	0.124	0.000	23.07	0.583	0.499	0.019
	Mult. Knaps.	0.00	0.000	0.663	0.000	0.00	0.000	4.096	0.151
mid	Rnd. Act.	52.57	0.840	0.802	0.130	58.04	4.708	2.889	0.732
	Rnd. Order	5.92	0.117	0.304	0.048	2.74	0.569	2.149	0.411
	Knapsack	58.94	0.195	0.134	0.027	28.70	0.536	0.442	0.089
	Mult. Knaps.	0.00	0.000	0.730	0.121	0.00	0.000	4.259	0.834

[a] Percentage of load-controlled cells exceeding load limit

[b/c] Average difference of cell power to $\bar{p}^{\downarrow}_{\max}$ of load-controlled cells exceeding/observing load limit [W]

[d] Average absolute difference of power values for perfect and discrete load control [W]

Table B.5: Performance of discrete load-control methods for Berlin network: fraction of served traffic in load-controlled cells

		speech service			all services		
intensity	method	% hi[a]	∅ ov[b]	∅ und[c]	% hi[a]	∅ ov[b]	∅ und[c]
low	Random Activation	57.14	0.011	0.014	62.68	0.053	0.078
	Random Order	10.71	0.014	0.010	12.13	0.027	0.063
	Knapsack	32.14	0.012	0.011	25.06	0.029	0.037
	Multiple Knapsack	0.00	0.000	0.022	1.68	0.008	0.109
mid	Random Activation	52.69	0.020	0.023	57.70	0.063	0.087
	Random Order	30.47	0.010	0.014	18.13	0.030	0.062
	Knapsack	55.81	0.032	0.020	33.51	0.040	0.040
	Multiple Knapsack	12.24	0.005	0.017	5.69	0.011	0.082

[a] Percentage of load-controlled cells serving more traffic than expected
[b] Average over-service (percentage points)
[c] Average under-service (percentage points)

Table B.6: Parameters and interference ratio for benchmark configurations

	site dist. [m]	tilt [°]	intf. ratio [%]
Berlin	1343	8	66.3
Lisbon	870	6	52.4
Turin	1173	4	135.7
Vienna	1550	8	141.4

Table B.7: Complete network score cards resulting from simplified Monte Carlo simulation

		coverage				pilot quality		downlink performance				uplink perf.		
		$A^{(\mathrm{F_c})}$	$A^{(\mathrm{F_c}/\mathrm{I_0})}\vert$	$\vert A^{(\mathrm{C})}\vert$	$\vert A^{(\mathrm{C})}\vert_{\tilde{\ell}}$	$A^{(\mathrm{SHO})}$	$A^{(\mathrm{PP})}\vert$	$\sum_i \bar{p}_{\bar{i}}$[W]	$\bar{\mathrm{L}}^{\downarrow}$	$\bar{\lambda}^{\downarrow}$	$\bar{\iota}^{\downarrow}$	$\bar{\mathrm{L}}^{\uparrow}$	$\bar{\lambda}^{\uparrow}$	$\bar{\iota}^{\uparrow}$
Berlin	prep.	100.0	98.6	98.6	98.6	23.9	1.1	773.2	35.9	94.0	114.7	16.3	100.0	66.8
	1-opt	100.0	99.8	99.8	99.8	22.9	1.0	660.7	30.7	97.9	91.2	15.1	100.0	54.5
	MIP	100.0	99.8	99.8	99.8	22.9	1.0	647.9	30.1	98.1	89.2	15.0	100.0	53.1
Lisbon	prep.	98.4	99.9	98.3	98.7	23.8	0.9	469.8	24.5	97.5	81.8	11.1	100.0	54.5
	1-opt	98.7	100.0	98.6	99.1	22.0	0.8	422.9	22.1	99.0	65.7	10.4	100.0	44.8
	MIP	98.7	100.0	98.7	99.1	22.2	0.7	422.6	22.1	99.0	65.8	10.4	100.0	44.9
Turin	prep.	84.2	99.6	83.8	64.9	20.7	1.0	1999.4	29.9	99.0	209.0	35.1	91.4	61.6
	1-opt	84.2	99.6	83.8	64.9	20.7	1.0	1995.6	29.9	99.0	208.9	35.1	91.4	61.4
	MIP	84.1	99.6	83.7	64.6	20.4	0.9	1838.7	27.5	99.4	196.1	34.1	92.1	55.1
Vienna	prep.	71.0	99.9	70.9	94.8	16.4	0.8	2257.1	31.3	98.4	191.6	11.1	100.0	79.5
	1-opt	70.9	99.9	70.8	94.7	16.2	0.8	2183.7	30.3	98.8	175.3	10.7	100.0	73.4
	MIP	70.9	99.9	70.8	94.7	16.2	0.8	2173.6	30.2	98.9	173.0	10.6	100.0	72.7

All units in percent unless stated otherwise. For the symbols used for performance indicators and their definitions see Sec. 3.4.

Acronyms and Symbols

Notation that is not listed here is defined where it is used. Plain numbers refer to pages of introduction, bracketed numbers to definitions.

Acronyms

2G/3G	2nd/3rd generation (telecommunications system)
BLER	block error rate
CDMA	code division multiple access, 11
CIR	carrier-to-interference ratio, 15
C.O.V.	coefficient of variation, 71
CPICH	common pilot channel, 14
dB	decibel (dimensionless), 11
dBm	decibel referenced to one milliwatt (power unit), 11
DL	downlink direction
DSL	digital subscriber line
E_b	received bit energy, 15
E_c	received chip energy, 14
FDD	frequency division duplex, 12
GPRS	general packet radio service
GSM	global system for mobile communications
HSPA	high speed packet access
I_0	interference and noise spectral density, 14
MIP	mixed integer (linear) program
N_0	interference and noise spectral density (including orthogonality), 15
OFDM	orthogonal frequency division multiplexing
RNC	radio network controller, 8
RRM	radio resource management
TCP	transmission control protocol
UL	uplink direction
UMTS	universal mobile telecommunications system
UTRA	UMTS terrestrial radio access
W-CDMA	wideband code division multiple access, 11
WLAN	wireless local area network
WWW	world wide web
XML	extensible markup language

Symbols

$.^{\uparrow}/.^{\downarrow}$	quantity defined in uplink/downlink
$x, \boldsymbol{x}, \boldsymbol{X}$	scalar, vector, matrix, 25
$\lvert\cdot\rvert_{\ell}^{\downarrow}$	area weighted by normalized user load intensity, (3.48b)
$\mathbf{1}_B$	indicator function of set B, 59
α_m	activity factor for user m, 21
A	complete scenario area, 19
A_i	best server area of cell i, (4.1)
$A^{(\mathrm{C})}$	covered area, (3.51)
$A^{(\mathrm{E_c})}$	$\mathrm{E_c}$-covered area, (3.49)
$A^{(\mathrm{E_c/I_0})}$	$\mathrm{E_c/I_0}$-covered area, (3.50)
$A^{(\mathrm{PP})}$	pilot-polluted area, (3.53)
$A^{(\mathrm{SHO})}$	soft handover area, (3.52)
β	weight for combining objective approximations, 119
C_i	coverage set of configuration i, (5.2)
$c_{ij}^{\uparrow}$	uplink interference coupling coefficient between cells i and j, (3.3)
$c_{ij}^{\downarrow}$	downlink interference coupling coefficient between cells i and j, (3.11)
$\boldsymbol{C}^{\uparrow}/\boldsymbol{C}^{\downarrow}$	uplink/downlink interference coupling matrix, 38/40
$\mathcal{D}_{ij}$	dominance set of i over j, (5.9)
$\mathrm{diag}(\boldsymbol{x})$	diagonal matrix with the elements of the vector $\boldsymbol{x}$ on the diagonal
$\mathcal{F}$	feasible network designs, 111
$f^{(\mathrm{C})}(\boldsymbol{z})$	coverage level of network design $\boldsymbol{z}$, (5.7)
$\gamma_i^{\uparrow}(\cdot)/\gamma_i^{\downarrow}(\cdot)$	path loss component of channel gain to/from cell i, $A \to [0,1]$, 20
$\tilde{\gamma}(\cdot)$	shadowing component of channel gain, $A \to \mathbb{R}_{>0}$, 20
$\gamma_{mi}^{\uparrow}/\gamma_{im}^{\downarrow}$	end-to-end channel gain between mobile m and cell i, 20
$\boldsymbol{I}$	identity matrix of matching dimension, 25
$\mathcal{I}/\mathcal{I}_s$	set of all potential antenna configurations/configurations for sector s, 111
$\bar{\iota}_i^{\downarrow}/\bar{\iota}^{\downarrow}$	average other-to-own-cell interference ratio in cell i/in planning area, (3.37)/(3.47b)
$\iota_i^{\uparrow}$	other-to-own-cell interference ratio at cell i
$\bar{\iota}^{\uparrow}$	average other-to-own-cell interference ratio in uplink, (3.47a)
$K^{(\mathrm{cov})}$	reference coverage value, 120
$\mathrm{L}_i/\bar{\mathrm{L}}$	cell load/average cell load, (3.41)–(3.45)
$\ell_m^{\uparrow}/\ell_m^{\downarrow}$	uplink/downlink user loading factor for user m, (3.2)/(3.9)
λ_i	traffic scaling factor/grade of service for cell i, 44
$\bar{\lambda}$	total grade of service (network aggregate), (3.46)
μ_m	CIR target for user m, 20
M/M_i	all mobiles/mobiles requesting service by cell i, 19
$\mu_{\mathrm{E_c/I_0}}$	$\mathrm{E_c/I_0}$ coverage threshold, 24
ν_i	price of selecting configuration i, 112

$\mathcal{N}$	set of cells, 19
$\eta_m^{\downarrow}/\eta^{\downarrow}(\cdot)$	noise at mobile m/noise function, Def. 4, 22
$\eta_i^{\uparrow}$	noise at cell i, 21
n	number of cells, 19
$N_{\max}$	limit for network cost, 119
$\mathrm{NR}_i^{\uparrow}$	noise rise at cell i, (2.5)
$\overline{\mathrm{NR}}^{\uparrow}$	average noise rise, (3.42)
$\omega(\cdot)/\omega_m$	orthogonality loss function/factor for mobile m, Def. 4, 21
$\mathcal{P}$	subdivision of the planning area ("pixels"), 113, 116
$p_i^{(\mathrm{C})}$	Total common channel power of cell i, 22
$p_i^{(\mathrm{P})}$	Pilot power of cell i, 24
$\bar{p}_i^{\downarrow}$	average total transmit power of cell i, (2.4)
$p_{im}^{\downarrow}$	Power emitted by cell i on the link to mobile m
$p_{\max}^{\downarrow}$	maximum nominal output power for cell i, 23
$\bar{p}_{\max}^{\downarrow}$	maximum average output power for cell i, 23
$p_i^{(\eta)}$	downlink noise load of cell i, (3.12)
$\pi_{\mathrm{E_c}}$	$\mathrm{E_c}$ coverage threshold, 24
$\boldsymbol{\psi}$	vector of capacity limits, 119
$p_{\max}^{(\mathrm{UE})}$	maximum output power of mobile m, 23
$\bar{p}_{\max}^{\uparrow}$	maximum average received power at cell i
$\bar{p}_i^{\uparrow}$	average total received power at cell i, (2.2)
$p_m^{\uparrow}$	power emitted by mobile m in uplink, 21
$\mathbb{R}_+/\mathbb{R}_{>0}$	Nonnegative/positive real numbers
$\rho(\mathbf{X})$	spectral radius of matrix $\mathbf{X}$ (largest modulus of an eigenvalue)
$\tau_i^{\downarrow}$	downlink traffic loading factor of cell i, (3.15)
T	user intensity function, $A \to \mathbb{R}_+$, 29
$T_\ell^{\downarrow}/T_\ell^{\uparrow}$	normalized uplink/downlink user load intensity function, (4.2)
u_{ia}	service variable for pixel a and configuration i, 116
$v_{ia}^{(j)}$	interference variable for server i, pixel a, and interferer j, 116
$\varsigma_{\min}^{(\mathrm{E_c})}$	minimum fraction of reference coverage to be attained, 120
y_a	coverage variable for pixel a, 113
z_i	selection variable for antenna configuration i, 111

Index

Bibliography

3G.co.uk. *Declines in mobile voice* ARPU *will continue to 2010 in western europe.* Press release, 2007. Retrieved on October 17, 2007 from http://www.3g.co.uk/PR/Sept2007/5122.htm.

3GToday.com. 3G *subscribers.* Website, 2007. Retrieved October 17, 2007 from http://www.3gtoday.com/wps/portal/subscribers; Source: Wireless Intelligence.

3GPP TSG RAN. *RF system scenarios. Tech Rep 3*GPP TR *25.942*, 3GPP, 2006*a*. v 2.1.3.

3GPP TSG RAN. *User equipment* (UE) *conformance specification; radio transmission and reception* (FDD)*; part 1: Conformance specification. Tech Rep 3*GPP TS *34.121-1*, 3GPP, 2006*b*. v 7.3.4, Retrieved April 14, 2007 from http://www.3gpp.org/ftp/Specs/html-info/34121-1.htm.

Actix Inc. *Actix radioplan—WiNeS.* Product brochure, 2006. Retrieved August 20, 2007 from http://www.actix.com/docs/pdf/radioplan.pdf.

Aircom International ltd. ADVANTAGE*—automatic cell planning and optimisation tool.* Product brochure, 2007. Retrieved August 20, 2007 from http://www.aircominternational.com/Brochures/advantage.pdf.

Altman Z, Picard JM, Jamaa SB, Fourestie B, Caminada A, Dony T, Morlier JF, Mourniac S. *Oasys: FT R&D* UMTS *automatic cell planning tool.* In *IEEE Int Symp AP-S*, pp 338–341. San Antonio, USA, 2002.

Amaldi E, Belotti P, Capone A, Malucelli F. *Optimizing base station location and configuration in* UMTS *networks.* Ann Oper Res, 146, pp 135–151, 2006.

Amaldi E, Capone A, Malucelli F. *Planning* UMTS *base station location: Optimization models with power control and algorithms.* IEEE Trans Wirel Comm, 2(5), pp 939–952, 2003.

ATIS Committee T1A1. ATIS *telecom glossary 2000. Tech Rep T1.523-2001*, Alliance for Telecommunications Industry Solutions/American National Standards Institute, Inc., 2001. Retrieved December 15, 2007 from http://www.atis.org/tg2k/.

Avriel M. *Nonlinear Programming: Analysis and Methods.* Prentice-Hall, 1976.

AWE Communications GmbH. *WinProp:* 3G UMTS CS. Product brochure, 2005*a*. Retrieved August 20, 2007 from http://www.awe-communications.de/Brochures/Brochure3G_CS.pdf.

AWE Communications GmbH. *WinProp: 3G UMTS PS*. Product brochure, 2005*b*. Retrieved August 20, 2007 from http://www.awe-communications.de/Brochures/Brochure3G_PS.pdf.

Balzanelli C, Munna A, Verdone R. W-CDMA *downlink capacity: Part 1*. In *Proc IEEE PIMRC'03*. Beijing, China, 2003.

Berman A, Plemmons RJ. *Nonnegative Matrices in the Mathematical Sciences*. Society for Industrial & Applied Mathematics, Philadelphia, revised edn, 1994.

Blum C, Roli A. *Metaheuristics in combinatorial optimization: Overview and conceptual comparison*. ACM Comput Surv, 35(3), pp 268–308, 2003.

Bottomley GE, Cairns DA, Cozzo C, Fulghum TL, Khayrallah AS, Lindell P, Sundelin M, Wang YPE. *Advanced receivers for* W-CDMA *terminal platforms and base stations*. Ericsson Rev, 2, pp 54–58, 2006.

Brady PT. *A model for generating on-off speech patterns in two-way conversation*. Bell System Tech J, 48, pp 2445–2472, 1969.

Buddendick H. *An Introduction to Static 3G Network Simulations within ProMan*. AWE Communications GmbH, 2005. Retrieved October 10, 2007 from http://www.awe-com.de/ApplicationNotes/StaticSimulation3G.pdf.

Burger S, Buddendick H, Wölfle G, Wertz P. *Location dependent* CDMA *orthogonality in system level simulations*. In *Proc IEEE VTC 2005 Spring*. Stockholm, Sweden, 2005.

Catrein D, Feiten A, Mathar R. *Uplink interference based call admission control for* W-CDMA *mobile communication systems*. In *Proc IEEE VTC 2005 Spring*. IEEE, Stockholm, Sweden, 2005.

Catrein D, Imhof L, Mathar R. *Power control, capacity, and duality of up- and downlink in cellular* CDMA *systems*. IEEE Trans Comm, 52(10), pp 1777–1785, 2004.

Chin FYL, Zhang Y, Zhu H. *Online* OSVF *code assignment with resource augmentation*. In *Proc. AAIM 2007*, pp 191–200. Portland, OR, 2007.

Chung SJ. *NP-completeness of the linear complementarity problem*. J Optim Theory Appl, 60(3), pp 393–400, 1989.

Chvátal V. *Linear Programming*. W. H. Freeman and Company, 1983.

Coello C. *An updated survey of GA-based multiobjective optimization techniques*. ACM Comput Surveys, 32(2), pp 109–143, 2000.

Cormen TH, Leiserson CE, Rivest RL. *Introduction to Algorithms*. MIT Press, Cambridge, MA, 1990.

Cosiro GmbH. *FUN radio network planning software for* UMTS. Product brochure, 2006. Retrieved August 20, 2007 from http://www.cosiro.com/downloads/FUN_Network_Type_UMTS_E.pdf.

De Schutter B, De Moor B. *The extended linear complementarity problem*. Math Programming, 71(3), pp 289–325, 1995.

De Schutter B, Heemels WP, Bemporad A. *On the equivalence of linear complementarity problems.* Oper Res Lett, 30(4), pp 211–222, 2002.

Deb K. *Multi-Objective Optimization using Evolutionary Algorithms.* John Wiley & Sons, Ltd, 2001.

Dehghan S. *A new approach.* 3GSM Daily 2005, 1, p 44, 2005.

Dehghan S, Lister D, Owen R, Jones P. W-CDMA *capacity and planning issues.* Electron & Comm Eng J, 12(3), pp 101–118, 2000.

Dolan E, Fourer R, Moré J, Munson T. *Optimization on the* NEOS *server.* SIAM News, 35(6), pp 4–9, 2002. The server address is http://neos.mcs.anl.gov/.

Doriga M, Stützle T. *Ant Colony Optimization.* MIT Press, 2004.

Du L, Biaham J, Cuthbert L. *A bubble oscillation algorithm for distributed geographic load balancing in mobile networks.* In *Proc IEEE INFOCOM 2004*, vol. 1. 2004.

Dziong Z, Krishnan M, Kumar S, Nanda S. *Statistical "snap-shot" for multi-cell* CDMA *system capacity analysis.* In *Proc IEEE WCNC 1999*, vol. 3, pp 1234–1237. 1999.

Eckhardt R. *Stan Ulam, John von Neumann, and the Monte Carlo method.* Los Alamos Science, 15, pp 131–137, 1987. Special Issue on Stanislaw Ulam 1909–1984.

Ehrgott M. *Multicriteria Optimization.* Springer, 2nd edn, 2005.

Ehrgott M, Gandibleux X. *Multiobjective combinatorial optimization—theory, methodology, and applications.* In Ehrgott M, Gandibleux X (eds), *Multiple Criteria Optimization: State of the Art Annotated Bibliographic Surveys*, chap. 8, pp 369–444. Kluwer Academic Publishers, Boston/Dodrecht/London, 2002.

Eisenblätter A, Fügenschuh A, Geerdes HF, Junglas D, Koch T, Martin A. *Optimization methods for* UMTS *radio network planning.* In *Proc Int. Conf. on Operations Research 2003*. Heidelberg, Germany, 2003.

Eisenblätter A, Fügenschuh A, Geerdes HF, Junglas D, Koch T, Martin A. *Integer programming methods for* UMTS *radio network planning.* In *Proc WiOpt'04*. Cambridge, UK, 2004.

Eisenblätter A, Geerdes HF. *A novel view on cell coverage and coupling for* UMTS *radio network evaluation and design.* In *Proc INOC'05*. ENOG, Lisbon, Portugal, 2005.

Eisenblätter A, Geerdes HF, Koch T, Martin A, Wessäly R. UMTS *radio network evaluation and optimization beyond snapshots.* Math Methods Oper Res, 63, pp 1–29, 2006.

Eisenblätter A, Geerdes HF, Munna A, Verdone R. *Comparison of models for* W-CDMA *downlink capacity assessment based on a* MORANS *reference scenario.* In *Proc IEEE VTC 2005 Spring*. IEEE, Stockholm, Sweden, 2005*a*.

Eisenblätter A, Geerdes HF, Rochau N. *Analytical approximate load control in* W-CDMA *radio networks.* In *Proc IEEE VTC 2005 Fall*. IEEE, Dallas, TX, 2005*b*.

Eisenblätter A, Koch T, Martin A, Achterberg T, Fügenschuh A, Koster A, Wegel O, Wessäly R. *Modelling feasible network configurations for* UMTS. In Anandalingam

G, Raghavan S (eds), *Telecommunications Network Design and Management*, pp 1–22. Kluwer, 2002.

Eisenblätter A. *Frequency Assignment in GSM Networks: Models, Heuristics, and Lower Bounds*. PhD thesis, TU Berlin, Göttingen, 2001.

Eisenblätter A, Geerdes HF, Türke U, Koch T. MOMENTUM *data scenarios for radio network planning and simulation (extended abstract)*. In *Proc WiOpt'04*. Cambridge, UK, 2004.

Eisenblätter A, Geerdes HF, Türke U. *Public UMTS radio network evaluation and planning scenarios*. Internat J Mobile Network Des and Innovation, 1(1), pp 40–53, 2005.

Ellinger F, Barras D, Jäckel H. *Mobile phone market study*. In *Proc Wireless 2002*. Calgary, Canada, 2002.

Ericsson AB. *Planet EV—the evolution of network design*. Product brochure, 2006. Retrieved August 20, 2007 from http://www.ericsson.com/solutions/tems/network_plan/downloads/planetev_ds.pdf.

Ericsson AB. *TEMS™ cellplanner—the optimizer's choice*. Product brochure, 2007. Retrieved August 20, 2007 from http://www.ericsson.com/solutions/tems/network_plan/downloads/TEMS_CellPlanner_7.1.pdf.

Feiten A. *Capacity Allocation and Resource Allocation in Wireless Communication Systems*. PhD thesis, RWTH Aachen, 2006.

Feiten A, Mathar R. *Optimal power control for multiuser CDMA channels*. In *Proc IEEE ISIT 2005*. Adelaide, Australia, 2005.

Fejes Tóth L. *Lagerungen in der Ebene, auf der Kugel und im Raum*. Springer, 1953.

Fischer C. *On CDMA load and the Perron root*. In *Proc. IZS 2006*, pp 126–129. 2006.

Fishman G, Fishman G. *Monte Carlo: Concepts, Algorithms and Applications*. Springer, 1996.

Fleuren MJ, Stüben H, Zegwaard GF. *MoDySim—A parallel dynamic UMTS simulator*. In *Proc PARCO 2003*, pp 347–354. Dresden, Germany, 2003.

Forsk SA. *Atoll—global RF planning solution*. Product brochure, 2007. Retrieved August 20, 2007 from http://www.forsk.com/images/product/atoll_overview/PlaqAtoll2_6.zip.

Galota M, Glaßer C, Reith S, Vollmer H. *A polynomial-time approximation scheme for base station positioning in UMTS networks*. In *Proc DIALM 2001*, pp 52–59. ACM Press, New York, NY, USA, 2001.

Geerdes HF. *Dynamic aspects in W-CDMA: HSPA performance*. STSM *scientific report*, COST 293 (GRAAL), 2007.

Geerdes HF, Eisenblätter A. *Reconciling theory and practice: A revised pole equation for W-CDMA cell powers*. In *Proc MSWIM'07*. Chania, Greece, 2007.

Geerdes HF, Ryll F. *Efficient approximation of blocking rates in* UMTS *radio networks*. In *Proc PGTS'06*. Wrocław, 2006.

Gerdenitsch A. *System Capacity Optimization of* UMTS FDD *Networks*. PhD thesis, Technische Universität Wien, 2004.

Gerdenitsch A, Jakl S, Chong YY, Toeltsch M. *A rule-based algorithm for common pilot channel and antenna tilt optimization in* UMTS FDD *networks*. ETRI J, 26(5), pp 437–442, 2004.

Gerdenitsch A, Jakl S, Toeltsch M, Neubauer T. *Intelligent algorithms for system capacity optimization of* UMTS FDD *networks*. In *Proc* IEE *3G2003*, pp 222–226. London, 2003.

Gil F, Claro AR, Ferreira JM, Pardelinha C, Correia LM. *A 3-D interpolation method for base-station antenna radiation patterns*. IEEE Antennas and Propagation Magazine, 43(2), pp 132–137, 2001.

Glaßer C, Reith S, Vollmer H. *The complexity of base station positioning in cellular networks*. Discrete Appl Math, 148(1), pp 1–12, 2005.

Haimes YY, Lasdon LS, Wismer DA. *On bicriterion formulation of the problems of integrated system identification and system optimization*. IEEE Trans Systems Man Cybernet, 1(3), pp 296–297, 1971.

Hanly S. *Congestion measures in* DS-CDMA *networks*. IEEE Trans Comm, 47(3), pp 426–437, 1999.

Hanly S, Mathar R. *On the optimal base station density for* CDMA *cellular networks*. IEEE Trans Comm, 50(8), pp 1274–1281, 2002.

Hedberg T, Parkvall S. *Evolving* W-CDMA. Ericsson Rev, 2, pp 124–131, 2000.

Heuser H. *Lehrbuch der Analysis*, vol. 2. B.G.Teubner, 9th edn, 1995.

Hills A. *Large-scale wireless* LAN *design*. IEEE Comm Magazine, 39(11), pp 98–107, 2001.

Holma H, Toskala A (eds). W-CDMA *for* UMTS . Jon Wiley & Sons, revised edn, 2001.

Holma H, Toskala A (eds). *HSDPA/HSUPA for* UMTS*: High Speed Radio Access for Mobile Communications*. John Wiley & Sons, 2006.

Hoshyar R, Bahai A, Tafazolli R. *Static simulation, solution of* UMTS *capacity and coverage load equations*. In *Proc PIMRC 2003*, vol. 1, pp 346–350. 2003.

Hoßfeld T, Mäder A, Staehle D. *When do we need rate control for dedicated channels in* UMTS*?* In *Proc IEEE VTC 2006 Spring*. Melbourne, Australia, 2006.

Hunukumbure M, Beach M, Allen B. *Downlink orthogonality factor in* UTRA FDD *systems*. Electron Lett, 38(4), pp 196–197, 2002.

Hurley S. *Planning effective cellular mobile radio networks*. IEEE Trans Veh Technol, 12(5), pp 243–253, 2002.

ILOG S. A. ILOG CPLEX *10.1—User's Manual*, 2006.

Imhof L, Mathar R. *The geometry of the capacity region for* CDMA *systems with general power constraints*. IEEE Trans Comm, 4(5), pp 2040–2044, 2005.

Isotalo T, Niemelä J, Lempiäinen J. *Electrical antenna downtilt in* UMTS *network*. In *Proc EW 2004*, pp 265–271. Barcelona, 2004.

Itô K (ed). *Encyclopedic Dictionary of Mathematics*, vol. III. MIT Press, 3rd edn, 1987. Entry "Monte Carlo simulation".

Jain R. *The Art of Computer Systems Performance Analysis*. John Wiley & Sons, 1991.

Jamaa SB, Altman Z, Picard JM, Fourestie B. *Multi-objective strategies for automatic cell planning of* UMTS *networks*. In *Proc IEEE VTC 2004 Fall*. Los Angeles, USA, 2004.

Jamaa SB, Altman Z, Picard JM, Fourestie B, Mourlon J. *Manual and automatic design for* UMTS *networks*. In *Proc WiOpt'03*. INRIA Press, Sophia Antipolis, France, 2003.

Jamaa SB, Altman Z, Picard JM, Ortega A. *Steered optimization strategy for automatic cell planning of* UMTS *networks*. In *Proc IEEE VTC 2005 Spring*. IEEE, Stockholm, Sweden, 2005.

Jedidi A, Caminada A, Finke G. *2-objective optimization of cells overlap and geometry with evolutionary algorithms*. In *Proc EvoWorkshops 2004*, vol. 3005 of *Lecture Notes in Computer Science*, pp 130–139. Springer, Coimbra, Portugal, 2004.

Jedidi A, Caminada A, Finke G, Dony T. *Geometrical constraint handling and evolutionary algorithm for cellular network design*. In *Proc INOC 2003*. Paris, France, 2003.

Johnson AS, Youssef S. *FreeHEP 2000*. In *Proc CHEP 2000*. Padova, Italy, 2000.

Karmarkar N. *A new polynomial-time algorithm for linear programming*. Combinatorica, 4(4), pp 373–395, 1984.

Kathrein catalogue. *790–220* MHz *Base Station Antennas for Mobile Communications*. Kathrein-Werke KG, Rosenheim, Germany, 2001.

Klenke A. *Wahrscheinlichkeitstheorie*. Springer, 2006.

Koch T. *Rapid Mathematical Programming*. PhD thesis, TU Berlin, Germany, 2004. Available at http://www.zib.de/Publications/abstracts/ZR-04-58/, ZIMPL is available at http://www.zib.de/koch/zimpl.

Kürner T. *Analysis of cell assignment probability predictions*. *Tech Rep TD (98) 05*, COST 259, 1998.

Kürner T. *Propagation models for macro cells*. In *Digital Mobile Radio Towards Future Generation Systems*, COST 231 *Final Report*. 1999.

Laiho J, Wacker A, Novosad T (eds). *Radio Network Planning and Optimisation for* UMTS. John Wiley & Sons, 2002.

Lamers E, Perera R, Türke U, Görg C. *Short term dynamic system level simulation concepts for* UMTS *network planning*. In *Proc ITC 18*, pp 101–110. Berlin, Germany, 2003.

Lee J, Miller L. CDMA *Systems Engineering Handbook*. Artech House, 1998.

Leibnitz K. *Analytical Modeling of Power Control and its Impact on Wideband* CDMA *Capacity and Planning*. PhD thesis, Universität Würzburg, 2003.

Li D, Sun X. *Nonlinear Integer Programming*. Springer, 2006.

Lustig M, Schlüter N, Tenckhoff J, Weller A. *Planning* UMTS *communication networks with the software tool PegaPlan*. T-Systems Enterprise Services GmbH, 2004. Retrieved August 20, 2007 from http://www.t-mobile.de/downloads/pegaproducts/pegaplan/article_planning_umts.pdf.

Madras N. *Lectures on Monte Carlo Methods*. American Mathematical Society, 2002.

Majewski K, Türke U, Huang X, Bonk B. *Analytical cell load assessment in OFDM radio networks*. In *Proc. IEEE PIMRC'07*. 2007.

Mathar R. *Mathematical modeling, design, and optimization of mobile communiciation networks*. Jber Dt Math-Verein, 103(3), pp 101–114, 2001.

Mathar R, Niessen T. *Optimum positioning of base stations for cellular radio networks*. Wirel Networks, 6(4), pp 421–428, 2000.

Mathar R, Schmeink M. *Optimal base station positioning and channel assignment for 3G mobile networks by integer programming*. Ann Oper Res, 107(1–4), pp 225–236, 2001.

Melachrinoudis E, Rosyidi B. *Optimizing the design of a* CDMA *cellular system configuration with multiple criteria*. Ann Oper Res, 106, pp 307–329, 2001.

Mendo L, Hernando JM. *On dimension reduction for the power control problem*. IEEE Trans Comm, 49(2), pp 243–248, 2001.

MOMENTUM Project. MOMENTUM *public* UMTS *planning scenarios*. Avaliable online at http://momentum.zib.de/data.php, 2003. IST-2000-28088.

Munna A, Boqué SR, Verdone R. MORANS—*mobile radio access network reference scenarios*. In *Proc. WPMC'04*. Abano Terme, Italy, 2004.

Murphy C, Dillon M, Diwan A. *Performance gains using remote control antennas in* CDMA *and 1xEV-DO networks*. In *Proc IEEE VTC 2005 Fall*. Dallas, TX, USA, 2005.

Mäder A, Staehle D. *Analytic modelling of the* W-CDMA *downlink capacity in multi-service environments*. In *Proc 16th ITC SS*, pp 229–238. Antwerp, Belgium, 2004.

Mäder A, Wagner B, Hoßfeld T, Staehle D, Barth H. *Measurements in a laboratory* UMTS *network with time-varying loads and different admission control strategies*. In *Proc IPS-MoMe 2006*. Salzburg, Austria, 2006.

Nasreddine J, Pérez-Romero J, Sallent O, Agusti R, Lagrange X. *A proposal on frequency management methodologies for* CDMA *systems using cell coupling matrices*. In *Proc IEEE VTC 2006 Fall*. 2006.

Nawrocki M, Aghvami H, Dohler M (eds). *Understanding* UMTS *Radio Network Modelling, Planning and Automated Optimisation: Theory and Practice*. John Wiley & Sons, Ltd, 2006.

Nawrocki MJ, Dohler M, Aghvami AH. *On cost function analysis for antenna optimization in* UMTS *networks.* In *Proc IEEE PIMRC'05*. Berlin, Germany, 2005.

Nemhauser GL, Wolsey LA. *Integer and Combinatorial Optimization*. John Wiley & Sons, Ltd, 1988.

Neto ELA, Yacoub MD, Shinoda AA, Pellenz ME. *A comprehensive 3G link level simulator.* In *Proc SS '02*, p 381. IEEE Computer Society, Washington, DC, USA, 2002.

Niemelä J, Borkowski J, Lempiäinen J. *Performance of static* W-CDMA *simulator.* In *Proc IEEE WPMC'05*. Aalborg, Denmark, 2005.

Niemelä J, Lempiäinen J. *Impact of mechanical antenna downtilt on performance of* W-CDMA *cellular network.* In *Proc IEEE VTC 2004 Spring*, pp 2091–2095. IEEE, Milan, Italy, 2004.

Niemelä J, Limpiäinen J. *Impact of base station locations and antenna orientations on* UMTS *radio network capacity and coverage evolution.* In *Proc IEEE WPMC 2003*, vol. 2, pp 82–86. 2003.

ETSI. UMTS *terrestrial radio access (*UTRA*); concept evaluation, version 3.0.0. Tech Rep* UMTS *30.06*, ETSI, 1997.

Nordling N. *Peak/average mode and bit rates. Open Technical Paper TMS-04:000050 Uen*, Ericsson AB, 2005. Retrieved August 28, 2007 from http://www.ericsson.com/solutions/tems/library/tech_papers/cellplanner/Peak-AverageModeandBitRates.pdf.

O'Brien T. *Jakarta Commons Cookbook*. O'Reilly, 2004.

Olmos J, Ruiz S. *Chip level simulation of the downlink in* UTRA-FDD. In *Proc IEEE PIMRC 2000*, vol. 2, pp 1469–1473vol.2. 2000.

Optimi, Inc. *x-ACP*. Product brochure, 2006*a*. Retrieved August 20, 2007 from http://www.optimi.com/pdf/www/optimi_x_acp.pdf.

Optimi, Inc. *x-Simulator*. Product brochure, 2006*b*. Retrieved August 20, 2007 from http://www.optimi.com/pdf/www/optimi_x_simulator.pdf.

Osyczka A. *Multicriterion Optimization in Engineering with* FORTRAN *Programs*. Ellis Horwood Ltd, 1984.

Papadimitriou CH, Steiglitz K. *Combinatorial Optimization: Algorithms and Complexity*. Prentice-Hall, Inc., Englewood Cliffs, New Jersey, USA, 1982.

Passerini C, Falciasecca G, Bordoni F. *Correlation between delay-spread and orthogonality factor in urban environments.* Electron Lett, 37(6), pp 384–386, 2001.

Pedersen K, Mogensen P. *The downlink orthogonality factors influence on* W-CDMA *system performance.* In *Proc VTC Fall 2002*, vol. 4, pp 2061–2065. 2002.

Rappaport T. *Wireless Communications: Principles and Practice*. Prentice Hall Int, 2nd edn, 2002.

Reininger P, Caminada A. *Multicriteria design model for cellular network.* Ann Oper Res, 107, pp 251–265, 2001.

Resnick SI. *Adventures in Stochastic Processes.* Birkhäuser, Boston, 1992.

Ryll F. *Approximating Blocking Rates in* UMTS *Radio Networks.* Master's thesis, TU Berlin, 2006.

Saunders SR. *Antennas and Propagation for Wireless Communication Systems.* John Wiley & Sons, New York, NY, 1999.

Sazonov V. *Algebra of sets.* In Hazewinkel M (ed), *Encyclopaedia of Mathematics*, vol. 1. Kluwer, 1988.

Schema Ltd. *Correcting the approach to optimal cell planning for* W-CDMA *networks.* White paper, 2003*a*. Available upon request via http://www.schema.com/.

Schema Ltd. *Meeting the challenge of* UMTS *optimization and planning.* White paper, 2003*b*. Available upon request via http://www.schema.com/.

Schema Ltd. *Ultima optiplanner™—cell parameter planning and optimization for* UMTS *networks.* Product brochure, 2007. Retrieved August 20, 2007 from http://www.schema.com/img/pdf/UltimaOptiPlannerBrochure.pdf.

Schmeink M. *Optimierungsalgorithmen zur automatisierten Funknetzplanung.* PhD thesis, RWTH Aachen, 2005.

Scholz P. *Basic Antenna Principles for Mobile Communications.* Kathrein-Werke KG, Rosenheim, Germany, 2000. Retrieved April 14, 2007 from http://www.kathrein.de/de/mca/techn-infos/download/BasicAntenna.pdf.

Schrijver A. *Theory of Linear and Integer Programming.* John Wiley & Sons, Ltd, 1986.

Schrijver A. *Combinatorial Optimization: Polyhedra and Efficiency.* Algorithms and Combinatorics. Springer, Berlin/Heidelberg, 2004.

Siomina I. *Radio Network Planning and Resource Optimization: Mathematical Models and Algorithms for* UMTS, WLANs *and Ad Hoc Networks.* PhD thesis, Linköping University, 2007.

Siomina I, Värbrand P, Yuan D. *Automated optimization of service coverage and base station antenna configuration in* UMTS *networks.* IEEE Wirel Comm (Special Issue on 3G/4G/WLAN/WMAN Planning and Optimization), pp 16–325, 2006.

Siomina I, Yuan D. *Pilot power management in* W-CDMA *networks: coverage control with respect to traffic distribution.* In *Proc ACM MSWiM 2004*, pp 276–282. ACM Press, 2004.

Siomina I, Yuan D. *Soft handover overhead control in pilot power management in* W-CDMA *networks.* In *Proc IEEE VTC 2005 Spring*. Stockholm, Sweden, 2005.

Sipilä K, Honkasalo K, Laiho-Steffens J, Wacker A. *Estimation of capacity and required transmission power of* W-CDMA *downlink based on a downlink pole equation.* In *Proc. VTC Spring 2000*. Tokyo, Japan, 2000.

Sipilä K, Laiho-Steffens J, Wacker A, Jäsberg M. *Modeling the impact of fast power control on the WCDM uplink.* In *Proc. IEEE VTC 1999 Spring*. Houston, Texas, 1999.

Sobczyk A. *Impact of the traffic distribution variation on the* W-CDMA *uplink performance.* In *Proc URSI'05*. Poznań, Poland, 2005.

Staehle D. *Analytical Methods for* UMTS *Radio Network Planning*. PhD thesis, Universität Würzburg, Würzburg, Germany, 2004.

Staehle D, Leibnitz K, Heck K. *Fast prediction of the coverage area in* UMTS *networks.* In *Proc IEEE* GLOBECOM. Taipei, Taiwan, 2002.

Staehle D, Leibnitz K, Heck K, Tran-Gia P, Schröder B, Weller A. *Analytic approximation of the effective bandwidth for best-effort services in* UMTS *networks.* In *Proc IEEE VTC 2003 Spring*. Jeju, South Korea, 2003.

Stańczak S, Wiczanowski M, Boche H. *Resource Allocation in Wireless Networks*. LNCS. Springer, 2006.

Stern HP, Mahmoud SA, Wong KK. *A comprehensive model for voice activity in conversational speech—development and application to performance analysis of new-generation wireless communication systems.* Wirel Netw, 2(4), pp 359–367, 1996.

Symena Software & Consulting GmbH. *The value of automated optimization—case study: Using capesso™ blue to improve the performance of a* W-CDMA *network.* Press release, 2003. Retrieved April 7, 2007 from http://www.symena.com/Value_Automated_Optimization.pdf.

Symena Software & Consulting GmbH. CAPESSO™ *automated radio network optimization.* Product brochure, 2004. Retrieved August 20, 2007 from http://www.symena.com/CAPblue.pdf.

Tran-Gia P, Staehle D, Leibnitz K. *Source traffic modeling of wireless applications.* Int J Electron Commun (AEÜ), 55(1), pp 1–10, 2001.

Tutschku K. *Models and Algorithms for Demand-oriented Planning of Telecommunication Systems*. PhD thesis, Universität Würzburg, 1999.

Tutschku K, Tran-Gia P. *Spatial traffic estimation and characterization for mobile communication network design.* IEEE J Sel Area Comm, 16, 1998.

Türke U. *Efficient Methods for* W-CDMA *Radio Network Planning and Optimization*. PhD thesis, Universität Bremen, 2006.

Türke U, Koonert M. *Advanced site configuration techniques for automatic* UMTS *radio network design.* In *Proc IEEE VTC 2005 Spring*. Stockholm, Sweden, 2005.

UMTS Forum. *Happy new year for global 3G mobile industry as* UMTS/W-CDMA *subscriptions hit 100 million.* Press release, 2006. Retrieved Octover 17, 2007 from http://www.umts-forum.org/content/view/1438/110/.

UMTS Forum. 3G/UMTS *commercial deployments*. Website, 2007. Retrieved July 04, 2007 from http://www.umts-forum.org/content/view/2000/123/.

Veeravalli VV, Sendonaris A. *The coverage-capacity tradeoff in cellular* CDMA *systems.* IEEE Trans Veh Technol, 48(5), pp 1443–1450, 1999.

Verdone R, Buehler H. MORANS *white paper. Tech Rep TD(03)057,* COST 273, Barcelona, Spain, 2003.

Viterbi AJ. CDMA*: Principles of Spread Spectrum Communication.* Prentice Hall Ptr, Reading, MA, 1st edn, 1995.

Viterbi AM, Viterbi AJ. *Erlang capacity of a power controlled* CDMA *system.* IEEE J Sel Area Comm, 11(6), pp 892–900, 1993.

Värbrandt P, Yuan D. *A mathematical programming approach for pilot power optimization in* W-CDMA *networks.* In *Proc ATNAC 2003.* Melbourne, Australia, 2003.

Wacker A, Laiho-Steffens J, Sipila K, Jasberg M. *Static simulator for studying* W-CDMA *radio network planning issues.* In *Proc VTC-Spring 1999,* vol. 3, pp 2436–2440. 1999.

Walke B. *Mobilfunknetze und ihre Protokolle,* vol. 1. B. G. Teubner, 3rd edn, 2001.

Walz FM. *An engineerig approach: Hierarchical optimization criteria.* IEEE Trans Automatic Control, 12, 1967.

Weal P, Corne D, Murphy C. *Optimisation of* CDMA*-based mobile telephone networks: Algorithmic studies on real-world networks.* In *Proc PPSN IX,* Springer LNCS. 2006.

Winter T, Türke U, Lamers E, Perera R, Serrador A, Correia L. *Advanced simulation approach for integrated static and short-term dynamic* UMTS *performance evaluation. Tech Rep D27,* MOMENTUM IST-2000-28088, 2003.

Wolpert DH, Macready WG. *No free lunch theorems for optimization.* IEEE Trans Evolutionary Comput, 1(1), pp 67–82, 1997.

Zdunek R, Nawrocki MJ. *Improved modeling of highly loaded* UMTS *network with nonnegative constraints.* In *Proc IEEE PIMRC'06.* 2006.

GPSR Compliance

The European Union's (EU) General Product Safety Regulation (GPSR) is a set of rules that requires consumer products to be safe and our obligations to ensure this.

If you have any concerns about our products, you can contact us on ProductSafety@springernature.com

In case Publisher is established outside the EU, the EU authorized representative is:

Springer Nature Customer Service Center GmbH
Europaplatz 3
69115 Heidelberg, Germany

Zeitfracht Medien GmbH
Ferdinand-Jühlke-Straße 7
99095 Erfurt, Deutschland
produktsicherheit@kolibri360.de